北大社·"十四五"普通高等教育本科规划教材
高等院校机械类专业"互联网＋"创新规划教材

机电工程专业英语（第3版）

Technical English for Mechanical and Electrical Engineering
(The Third Edition)

主　编　杨春杰　王培玲
副主编　陈　琢　程思遥　丁晚景

内 容 简 介

本书共有32篇课文,主要内容包括机械设计基础理论与方法、金属材料及热加工、数控技术、机电一体化技术、汽车工程和科技写作等方面的专业英语知识。同时,为了反映本学科的发展趋势,增加了智能制造技术、无人驾驶汽车技术等内容。书后附有科技英语翻译及写作的简单介绍,以及参考译文。

本书可作为机械设计制造及自动化、智能制造工程、机械电子工程、车辆工程、机器人工程等本科专业及相关研究生专业的英语教材,也可作为从事机电工程工作的技术人员的参考用书。

图书在版编目(CIP)数据

机电工程专业英语 / 杨春杰,王培玲主编. -- 3版. -- 北京:北京大学出版社,2025.1. -- (高等院校机械类专业"互联网+"创新规划教材). -- ISBN 978-7-301-35512-1

Ⅰ.TH

中国国家版本馆CIP数据核字第2024EP4264号

书　　　名	机电工程专业英语(第3版) JIDIAN GONGCHENG ZHUANYE YINGYU (DI-SAN BAN)
著作责任者	杨春杰　王培玲　主编
策 划 编 辑	童君鑫
责 任 编 辑	关　英
数 字 编 辑	蒙俞材
标 准 书 号	ISBN 978-7-301-35512-1
出 版 发 行	北京大学出版社
地　　　址	北京市海淀区成府路205号　100871
网　　　址	http://www.pup.cn　新浪微博:@北京大学出版社
电 子 邮 箱	编辑部 pup6@pup.cn　总编室 zpup@pup.cn
电　　　话	邮购部 010-62752015　发行部 010-62750672　编辑部 010-62750667
印 刷 者	三河市北燕印装有限公司
经 销 者	新华书店
	787毫米×1092毫米　16开本　16.75印张　486千字 2006年4月第1版　2010年1月第2版 2025年1月第3版　2025年1月第1次印刷
定　　　价	49.80元

未经许可,不得以任何方式复制或抄袭本书之部分或全部内容。
版权所有,侵权必究
举报电话:010-62752024　电子邮箱:fd@pup.cn
图书如有印装质量问题,请与出版部联系,电话:010-62756370

Preface to the Third Edition

We have revised and supplemented the third edition of this book, incorporating valuable feedback from teachers and students of various universities who used the previous editions.

This book consists of 32 lessons, mainly covering professional English knowledge in the fields of basic theories and methods of mechanical design, metal materials heat processing, numerical control, mechatronics technology, automobile engineering, and scientific and technological writing. The report of the 20th National Congress of the Communist Party proposed to focus on the real economy in economic development, promote new industrialization, and accelerate the construction of a manufacturing power. This book adds content on topics such as manufacturing technology and unmanned driving technology. At the end of the book, there is also a brief introduction to the translation and writing of English for science and technology and reference translations, which is convenient for readers.

This book has certain professionalism and difficulty. Through in-depth study of this book, readers can improvetheir ability to read and translate professional English literature. This book could be used as a professional English textbook for undergraduate majors, such as mechanical design manufacturing and automation, intelligent manufacturing engineering, mechatronic engineering, vehicle engineering, and robot engineering, as well as related graduate majors. It also serves as a reference for technical personnel engaged in mechanical and electrical engineering work.

Edited by Yang Chunjie and Wang Peiling, who serve as the chief editors, with Chen Zhuo, Cheng Siyao, and Ding Wanjing as deputy editors, the book's content is divided among the editors as follows: Yang Chunjie wrote the chapters related to the basic theories and methods of mechanical design and scientific and technological writing; Wang Peiling wrote the chapters related to automotive engineering; Chen Zhuo wrote the chapters related to numerical control; Cheng Siyao wrote the chapters related to mechatronics technology; Ding Wanjing wrote the chapters related to metal materials and heat treatment.

Keeping pace with the information age, this book adds some video resources in the form of quick response codesnext to relevant knowledge points. Readers can scan the quick response codes to access more learning materials.

Due to the editors' limited time and ability, there are inevitably omissions and improper parts in the book, and we sincerely welcome readers' criticism and correction.

Editors

July 2024

第 3 版前言

本书第 3 版是编者在前两版的基础上，吸取多所高校师生在使用本书过程中提出的宝贵意见修订的。

本书共有 32 篇课文，主要内容包括机械设计基础理论与方法、金属材料及热加工、数控技术、机电一体化技术、汽车工程和科技写作等方面的专业英语知识。党的二十大报告提出，坚持把发展经济的着力点放在实体经济上，推进新型工业化，加快建设制造强国。本书增加了智能制造技术、无人驾驶汽车技术等内容。书后附有科技英语翻译及写作的简单介绍，以及参考译文，便于读者学习。

本书具有一定的专业性和难度，通过对本书的深入学习，读者可以提高阅读和翻译专业英语文献资料的能力。本书可作为机械设计制造及自动化、智能制造工程、机械电子工程、车辆工程、机器人工程等本科专业及相关研究生专业的英语教材，也可作为从事机电工程工作的技术人员的参考用书。

本书第 3 版由杨春杰、王培玲任主编，陈琢、程思遥、丁晚景任副主编。具体编写分工如下：杨春杰编写了机械设计基础理论与方法相关课文及科技论文写作相关内容；王培玲编写了汽车工程相关课文；陈琢编写了数控技术相关课文；程思遥编写了机电一体化技术相关课文；丁晚景编写了金属材料及热加工相关课文。

本书紧跟信息时代的步伐，采用"互联网＋"思维，以在相关知识点旁增加二维码的形式展示视频资源，读者可以扫描二维码来阅读更多学习资料。

由于编者时间和水平有限，书中难免有疏漏之处，敬请广大读者批评指正。

编　者
2024 年 7 月

【资源索引】

目 录

Lesson 1　Introduction to Mechanical Design 1
　1.1　Text 1
　1.2　Words and Phrases 4
　1.3　Complex Sentence Analysis 4
　1.4　Exercise 5

Lesson 2　Introduction to Engineering Mechanics—Fundamental Concepts 7
　2.1　Text 7
　2.2　Words and Phrases 10
　2.3　Complex Sentence Analysis 11
　2.4　Exercise 11

Lesson 3　Introduction to Fluid Mechanics—Surface Tension 13
　3.1　Text 13
　3.2　Words and Phrases 15
　3.3　Complex Sentence Analysis 16
　3.4　Exercise 16

Lesson 4　Machine Parts (Ⅰ) 18
　4.1　Text 18
　4.2　Words and Phrases 21
　4.3　Complex Sentence Analysis 22
　4.4　Exercise 22

Lesson 5　Machine Parts (Ⅱ) 24
　5.1　Text 24
　5.2　Words and Phrases 26
　5.3　Complex Sentence Analysis 28
　5.4　Exercise 28

Lesson 6　Mechanisms 29
　6.1　Text 29
　6.2　Words and Phrases 30
　6.3　Complex Sentence Analysis 31
　6.4　Exercise 32

Lesson 7　Fluid and Hydraulic System 33
7.1　Text 33
7.2　Words and Phrases 35
7.3　Complex Sentence Analysis 36
7.4　Exercise 36

Lesson 8　Engineering Graphic 38
8.1　Text 38
8.2　Words and Phrases 41
8.3　Complex Sentence Analysis 41
8.4　Exercise 42

Lesson 9　Introduction to CAD/CAM/CAPP 43
9.1　Text 43
9.2　Words and Phrases 45
9.3　Complex Sentence Analysis 45
9.4　Exercise 46

Lesson 10　Engineering Tolerance 48
10.1　Text 48
10.2　Words and Phrases 51
10.3　Complex Sentence Analysis 51
10.4　Exercise 52

Lesson 11　A Discussion on Modern Design Optimization 54
11.1　Text 54
11.2　Words and Phrases 57
11.3　Complex Sentence Analysis 58
11.4　Exercise 59

Lesson 12　Using Dynamic Simulation in the Development of Construction Machinery 60
12.1　Text 60
12.2　Words and Phrases 65
12.3　Complex Sentence Analysis 66
12.4　Exercise 66

Lesson 13　Introduction to Heat Pipe Technology in Machining Process 68
13.1　Text 68
13.2　Words and Phrases 70
13.3　Complex Sentence Analysis 71
13.4　Exercise 72

Lesson 14	Introduction to Material Forming	73
14.1	Text	73
14.2	Words and Phrases	76
14.3	Complex Sentence Analysis	77
14.4	Exercise	78

Lesson 15	Material Forming Processes	79
15.1	Text	79
15.2	Words and Phrases	83
15.3	Complex Sentence Analysis	84
15.4	Exercise	84

Lesson 16	Introduction to Mould	86
16.1	Text	86
16.2	Words and Phrases	89
16.3	Complex Sentence Analysis	90
16.4	Exercise	90

Lesson 17	Mould Design and Manufacturing	91
17.1	Text	91
17.2	Words and Phrases	93
17.3	Complex Sentence Analysis	94
17.4	Exercise	95

Lesson 18	Heat Treatment of Metal	96
18.1	Text	96
18.2	Words and Phrases	98
18.3	Complex Sentence Analysis	99
18.4	Exercise	100

Lesson 19	Numerical Control System	101
19.1	Text	101
19.2	Words and Phrases	103
19.3	Complex Sentence Analysis	104
19.4	Exercise	105

Lesson 20	Virtual Manufacturing	106
20.1	Text	106
20.2	Words and Phrases	108
20.3	Complex Sentence Analysis	108
20.4	Exercise	109

Lesson 21　Industrial Big Data for Decision-making in Intelligent Manufacturing ……… 110
   21.1　Text ……………………………………………………………………………… 110
   21.2　Words and Phrases …………………………………………………………… 112
   21.3　Complex Sentence Analysis ………………………………………………… 112
   21.4　Exercise ………………………………………………………………………… 113

Lesson 22　Robotics and Computer-integrated Manufacturing ……………………… 115
   22.1　Text ……………………………………………………………………………… 115
   22.2　Words and Phrases …………………………………………………………… 120
   22.3　Complex Sentence Analysis ………………………………………………… 121
   22.4　Exercise ………………………………………………………………………… 121

Lesson 23　Product Test and Quality Control …………………………………………… 123
   23.1　Text ……………………………………………………………………………… 123
   23.2　Words and Phrases …………………………………………………………… 125
   23.3　Complex Sentence Analysis ………………………………………………… 125
   23.4　Exercise ………………………………………………………………………… 126

Lesson 24　Mechatronics ………………………………………………………………… 127
   24.1　Text ……………………………………………………………………………… 127
   24.2　Words and Phrases …………………………………………………………… 129
   24.3　Complex Sentence Analysis ………………………………………………… 130
   24.4　Exercise ………………………………………………………………………… 130

Lesson 25　Introduction to MEMS ……………………………………………………… 132
   25.1　Text ……………………………………………………………………………… 132
   25.2　Words and Phrases …………………………………………………………… 134
   25.3　Complex Sentence Analysis ………………………………………………… 134
   25.4　Exercise ………………………………………………………………………… 135

Lesson 26　Industrial Robots …………………………………………………………… 136
   26.1　Text ……………………………………………………………………………… 136
   26.2　Words and Phrases …………………………………………………………… 139
   26.3　Complex Sentence Analysis ………………………………………………… 140
   26.4　Exercise ………………………………………………………………………… 140

Lesson 27　An Army of Small Robots ………………………………………………… 141
   27.1　Text ……………………………………………………………………………… 141
   27.2　Words and Phrases …………………………………………………………… 143
   27.3　Complex Sentence Analysis ………………………………………………… 144
   27.4　Exercise ………………………………………………………………………… 145

Lesson 28　Introduction to Automobile Engine ⋯⋯ 146
- 28.1　Text ⋯⋯ 146
- 28.2　Words and Phrases ⋯⋯ 149
- 28.3　Complex Sentence Analysis ⋯⋯ 150
- 28.4　Exercise ⋯⋯ 150

Lesson 29　Introduction to Automobile Chassis ⋯⋯ 152
- 29.1　Text ⋯⋯ 152
- 29.2　Words and Phrases ⋯⋯ 155
- 29.3　Complex Sentence Analysis ⋯⋯ 156
- 29.4　Exercise ⋯⋯ 156

Lesson 30　New Energy Vehicle ⋯⋯ 158
- 30.1　Text ⋯⋯ 158
- 30.2　Words and Phrases ⋯⋯ 161
- 30.3　Complex Sentence Analysis ⋯⋯ 161
- 30.4　Exercise ⋯⋯ 162

Lesson 31　Autonomous Vehicle ⋯⋯ 163
- 31.1　Text ⋯⋯ 163
- 31.2　Words and Phrases ⋯⋯ 165
- 31.3　Complex Sentence Analysis ⋯⋯ 166
- 31.4　Exercise ⋯⋯ 166

Lesson 32　How to Write a Scientific Paper ⋯⋯ 168
- 32.1　Text ⋯⋯ 168
- 32.2　Words and Phrases ⋯⋯ 171
- 32.3　Complex Sentence Analysis ⋯⋯ 172
- 32.4　Exercise ⋯⋯ 173

附录 A　关于科技英语翻译 ⋯⋯ 174

附录 B　科技论文的英文摘要 ⋯⋯ 175

附录 C　参考译文 ⋯⋯ 188

参考文献 ⋯⋯ 257

Lesson 1　Introduction to Mechanical Design

1.1　Text

【拓展视频】

[1]Machinery design is either to formulate an engineering plan to satisfy a specified need or to solve an engineering problem. It encompasses a range of disciplines, including materials, mechanics, heat, flow, control, and electronics.

Machinery design can be simple or complex, easy or difficult, mathematical or nonmathematical. It may involve a trivial problem or one of great importance. Good design is the orderly and interesting arrangement of an idea that provides specific results or effects. A well-designed product is functional, efficient, and dependable. Such a product is often less expensive than a similar, poorly designed product that does not function properly and requires constant repairs.

Individuals who perform the various functions of machinery design are typically called industrial designers. They must first carefully define the problem using an engineering approach to ensure that any proposed solutions effectively address the issue. [2]It is important for the designer to begin by identifying satisfactory solutions and distinguishing between them in order to identify the best option. Therefore, industrial designers must possess creative imagination as well as knowledge of engineering, production techniques, tools, machines, and materials to design a new product for manufacturing or to improve an existing product.

In the modern industrialized world, the wealth and living standards of a nation are closely linked to its capabilities in designing and manufacturing engineering products. It can be argued that advancements in machinery design and manufacturing can remarkably promote the overall level of a country's industrialization. Our country is playing an increasingly vital role in the global manufacturing industry. To accelerate this industrialization process, highly skilled design engineering with extensive knowledge and expertise are needed.

1. Machinery Components

The major part of a machine is the mechanical system. [3]The mechanical system is decomposed into mechanisms which can be further decomposed into mechanical components. In this sense, mechanical components are the

【拓展视频】

fundamental elements of machinery. Overall, mechanical components can be classified as universal and special components. Bolts, gears, and chains are typical examples of universal components, which can be used extensively in different machines across various industrial sectors. Turbine blades, crankshaft, and aircraft propeller are examples of special components, designed for specific purposes.

2. Mechanical Design Process

Product design requires extensive research and development. Many concepts of an idea must be studied, tested, refined, and then either used or discarded. Although the content of each engineering problem is unique, designers follow a similar process to solve these problems. The complete process is often outlined as Figure 1.1.

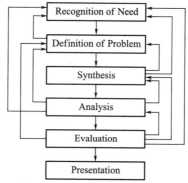

Figure 1.1 Design Process Model

(1) Recognition of Need.

Design sometimes begins when a designer recognizes a need and decides to address it. The need is often not immediately evident. Recognition is usually triggered by a particular adverse circumstance or a set of random circumstances that arise almost simultaneously. Identifying a need usually involves an undefined and vague problem statement.

(2) Definition of Problem.

Defining the problem is necessary to fully understand it, allowing the goal to be restated in a more reasonable and realistic way than the original problem statement. The problem definition must include all specifications for what is to be designed. Obvious items in the specifications are speeds, feeds, temperature limitations, maximum range, expected variation in the variables, and dimensional and weight limitations.

(3) Synthesis.

Synthesis involves seeking as many alternative design approaches as possible, usually without regard for their value or quality. This step is also sometimes called the ideation and invention phase, during which the largest possible number of creative solutions is generated. The synthesis activity includes specifying material, adding geometric features, and incorporating greater dimensional detail into the overall design.

(4) Analysis.

Analysis is a method of determining or describing the nature of something by breaking it down into its parts. In this process, the elements or nature of the design are analyzed to assess the fit between the proposed design and the original design goals.

(5) Evaluation.

Evaluation is the final proof of a successful design and usually involves testing a prototype in the laboratory. Here, we aim to discover if the design truly satisfies the needs.

The above description may give the erroneous impression that this process can be accomplished in a linear fashion as listed. On the contrary, iteration is required throughout the entire process, allowing movement from any step back to any previous step.

(6) Presentation.

Communicating the design to others is the final and vital presentation step in the design process. Basically, there are three means of communication: written, oral, and graphical. A successful engineer will be technically competent and versatile in all three forms of communication. A competent engineer should not fear the possibility of not succeeding in a presentation. In fact, the greatest gains are achieved by those willing to risk defeat.

3. Contents of Machinery Design

Machinery design is a fundamental technological course in mechanical engineering education. Its objective is to provide the concepts, procedures, data, and decision analysis techniques necessary to design machine elements commonly found in mechanical devices and systems; to develop engineering students' competence in machine design, which is the primary concern of machinery manufacturing and the key to produce quality products.

【拓展视频】

Machinery design covers the following contents:

(1) Provide an introduction to the design process, problem formulation, and safety factors.

(2) Review material properties and static and dynamic loading analysis, including beam, vibration, and impact loading.

(3) Review the fundamentals of stress and defection analysis.

(4) Introduce static failure theories and fracture mechanics analysis for static loads.

(5) Introduce fatigue-failure theory with an emphasis on stress-life approaches to high-cycle fatigue design, commonly used in the design of rotating machinery.

(6) Discuss thoroughly the phenomena of wear mechanisms, surface contact stresses, and surface fatigue.

(7) Investigate shaft design using fatigue-analysis techniques.

(8) Discuss fluid-film and rolling-element bearing theory and application.

(9) Provide a thorough introduction to the kinematics, design, and stress analysis of spur gears, and a simple introduction to helical, bevel, and worm gearing.

(10) Discuss spring design, including helical compression, extension, and torsion springs.

(11) Deal with screws and fasteners, including power screw and preload fasteners.

(12) Introduce the design and specification of disk and drum brakes.

1.2 Words and Phrases

machinery	n. [总称] 机器，机械
trivial	adj. 琐细的，平常的，微不足道的
mechanism	n. 机构
chain	n. 链（条），一连串，一系列
turbine blade	涡轮机叶片
crankshaft	n. 曲轴
propeller	n. 尤指轮船、飞机上的螺旋推进器
discard	v. 丢弃，抛弃，摒弃
recognition	n. 识别
trigger	v. 引发，引起，触发
vague	adj. 含糊的，不清楚的
synthesis	n. 综合
ideation	n. 构思能力，思维能力，构思过程
aggregate	adj. 合计的，集合的
prototype	n. 样机，原型
erroneous	adj. 错误的，不正确的
iteration	n. 反复
competent	adj. 有能力的，胜任的
versatile	adj. 通用的，万能的，多才多艺的

1.3 Complex Sentence Analysis

[1] Machinery design is either to formulate an engineering to satisfying a specified need or to solve an engineering problem.
① either…or…：或……或……
② formulate：明确地表达，认真阐明。

[2] It is important for the designer to begin by identifying satisfactory solutions and distinguishing between them in order to identify the best option.

① it 是形式主语，指代 to begin…，意为"对于设计师来说，……是很重要的"。
② by 为介词，表示方式，后接动名词 identifying 和 distinguishing；and 表示递进关系；in order to 表示目的。

[3] The mechanical system is decomposed into mechanisms which can be further decomposed into mechanical components.

① be decomposed into：被分解为。
② which 引导定语从句，在从句中作主语，指 mechanisms。

1.4 Exercise

1.4.1 Translate the following paragraph

The practice of design can be one of the most exciting and fulfilling activities for an engineer. There is great satisfaction and pride in seeing one's creative efforts turn into actual products and processes that benefit people. To excel in design requires several qualities. A design engineer should not only have solid technical training but also possess sound judgment and broad experience—qualities typically gained through significant professional work. A good start can be made with a capable teacher while still at university. However, beginning designers should expect to gain a substantial part of this training after graduation through further reading and study, and particularly through collaboration with other competent engineers. The more an engineer knows about all phrases of design, the better. Design is a demanding profession, and it is highly fascinating when practiced with a broad knowledge base.

1.4.2 Choose the proper answer to fill in the blanks and translate the sentences

1. They are using a (　　) shovel to clear up the streets.
2. Many products are made by (　　).
3. There is not a (　　) who hasn't had this problem.
4. Machinery design involves a range of disciplines in materials, (　　), heat, flow, control, electronics, and production.
5. The (　　) is worked by wind power.

A. machinery：n. machines or machine parts in general.
1. 机器，机械
2. 机械装置
3. 方法
4. 制造舞台效果的装置

5. 文学手段，（文学作品的）情节

B. mechanical：*adj*. of or pertaining to machines or tools.

1. 机械的，用机械的

2. 似机械的，呆板的，无表情（或感情）的，无意识的

3. 机械学的，力学的，物理的

C. machine：*n*. a system formed and connected to alter, transmit, and direct applied forces to accomplish a specific objective.

1. 机器，机械

2. 计算机

3. 汽车，自行车，飞机

4. 机构，操纵组织的核心集团

5. 机器人似的工作的人，没有感情或意志的人

D. mechanic：*n*. a worker skilled in making, using, or repairing machines.

机械工，修理工，技工

E. mechanics：*n*. the analysis of the action of forces on matter or material systems.

1. 力学，机械学

2. 技术性的部分，技术，技巧

Lesson 2   Introduction to Engineering Mechanics—Fundamental Concepts

2.1 Text

Before we begin our study of engineering mechanics, it is important to understand the meaning of certain fundamental concepts and principles.

1. Basic Quantities

The following four quantities are used throughout mechanics:

(1) Length. Length is used to locate the position of a point in space and thereby describe the size of a physical system. Once a standard unit of length is defined, one can then use multiples of this unit to define distances and geometric properties of a body.

(2) Time. Time is conceived as a succession of events. Although the principles of statics are time-independent, this quantity plays an important role in the study of dynamics.

(3) Mass. Mass is a measure of the quantity of matter that is used to compare the action of one body with that of another. Mass manifests itself as a gravitational attraction between two bodies and provides a measure of the resistance of matter to a change in velocity.

(4) Force. In general, force is considered as a "push" or "pull" exerted by one body on another. This interaction can occur when there is direct contact between the bodies, or it can occur when the bodies are physically separated by a small distance.

Examples include gravity, electric field force, and magnetic forces. In any case, a force is completely characterized by its magnitude, direction, and point of action.

2. Idealizations

Models or idealizations are used in mechanics to simplify the application of theories. The following section will introduce three important idealizations:

(1) Particles. A particle has mass, but its size can be neglected (Figure 2.1). For example, the size of the earth is insignificant compared to the size of its orbit, and therefore, the earth can be modeled as a particle when studying its orbital motion. When a body is idealized as a particle, the principles of mechanics are reduced to a rather simplified form since the geometry of the body will not be involved in the analysis of the problem.

(2) Rigid Body. [1] A rigid body can be considered as a combination of a large number of

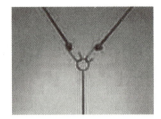

Figure 2.1 Particle

Three forces act on the ring. Since these forces all meet at a point,
we can assume the ring to be represented as a particle for any force analysis.

particles in which all the particles remain at a fixed distance from one another, both before and after applying a load. This model is important because the body's shape does not change when a load is applied, so we do not have to consider the type of material from which the body is made. In most cases, the actual deformations occurring in structures, machines, mechanisms, and the like are relatively small, making the rigid-body assumption suitable for analysis.

(3) Concentrated Force. A concentrated force represents the effect of a loading assumed to act at a point on a body. [2] We can represent a load by a concentrated force, provided the area over which the load is applied is very small compared to the overall size of the body. An example would be the contact force between a wheel and the ground (Figure 2.2).

Figure 2.2 Concentrated Force

Steel is a common engineering material that does not deform much under load.
Therefore, we can consider this railroad wheel to be a rigid body acted upon by the concentrated force of the rail.

3. Newton's Law of Motion

Engineering mechanics is formulated on the basis of Newton's three laws of motion, the validity of which is based on experimental observation. These laws apply to the motion of a particle as measured from a inertial (non-accelerating) reference frame. They may be briefly stated as follows:

(1) Newton's First Law.[3] A particle originally at rest or moving in a straight line with constant velocity tends to remain in this state provided the particle is not subjected to an unbalanced

force (Figure 2.3).

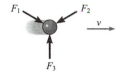

Figure 2.3 Equilibrium

(2) Newton's Second Law. A particle acted upon by an unbalanced force F experiences an acceleration a that has the same direction as the force and a magnitude that is directly proportional to the force F (Figure 2.4). If F is applied to a particle of mass m, this law may be expressed mathematically as $F=ma$.

Figure 2.4 Accelerated Motion

(3) Newton's Third Law. The mutual forces of action and reaction between two particles are equal, opposite, and collinear (Figure 2.5).

Figure 2.5 Action-Reaction

4. Newton's Law of Universal Gravitation

Shortly after formulating his three laws of motion, Newton postulated a law governing the gravitational attraction between any two particles. Stated mathematically:

$$F=G\frac{m_1 m_2}{r^2} \tag{2-1}$$

where, F—force of gravitation between the two particles;

G—universal constant of gravitation, according to experimental evidence, $G=6.673\times 10^{-11}$ m^3/(kg·s^2);

m_1, m_2—mass of each of the two particles;

r—distance between the two particles.

5. Weight

According to the Equation 2-1, any two particles or bodies have a mutual attractive (gravitational) force acting between them. For a particle located at or near the Earth's surface, the only significant gravitational force is between the Earth and the particle. This force, termed the weight, will be the only gravitational force considered in our study of mechanics.

From the Equation 2-1, we can develop an approximate expression for finding the weight W of a particle having a mass $m_1 = m$. Assuming the Earth to be a non-rotating sphere of constant density and having a mass $m_2 = M_e$, then if r is the distance between the Earth's center and the particle, we have:

$$F = G \frac{mM_e}{r^2}. \tag{2-2}$$

Letting $g = GM_e/r^2$:

$$W = mg.$$

By comparison with $F = ma$, we see that g is the acceleration due to gravity. Since it depends on r, the weight of a body is not an absolute quantity (Figure 2.6). Instead, its magnitude is determined by the location of measurement. For most engineering calculations, however, g is determined at sea level and at a latitude of 45°, considered the "standard location".

Figure 2.6　Weight

The astronaut's weight is diminished since she is far removed from the Earth's gravitational field.

2.2　Words and Phrases

mechanics	n. 力学，机械学，构成法，技术
geometric property	几何性质
statics	n. 静力学，静态
independent	adj. 独立的，公正的，无关联的；n. 无党派议员
dynamics	n. 动力学，相互作用，动态；驱动力
mass	n. 质量，大量，民众；adj. 一大群的；v. （使）聚集
particle	n. 微粒，颗粒；粒子，质点，极小量
rigid body	刚体，刚性体
equilibrium	n. 平衡，均势，平静
collinear	adj. 共线的，同线的

Introduction to Engineering Mechanics—Fundamental Concepts Lesson 2

mutual	*adj.* 共有的，共同的，相互的，彼此的
gravity	*n.* 重力，严重性，庄重

2.3 Complex Sentence Analysis

[1] A rigid body can be considered as a combination of a large number of particles in which all the particles remain at a fixed distance from one another, both before and after applying a load.

① 主句为"… can be considered as …"：……可以视为……
② in which 引导从句，解释质点之间保持固定距离。
③ both before and after …为状语结构，对上述位置关系作时间限定。

[2] We can represent a load by a concentrated force, provided the area over which the load is applied is very small compared to the overall size of the body.

① provided：前提是……，假如……，在……的条件下。
② compared to：与……比较。

[3] A particle originally at rest, or moving in a straight line with constant velocity, tends to remain in this state provided the particle is *not* subjected to an unbalanced force（Figure 2.3）.

① 主句为"A particle tends to remain in this state"：质点倾向于保持这种状态。
② 波浪线部分是对主语"A particle"的限定。
③ 虚线部分由 provided 引导从句，对主句进行前提限定。

2.4 Exercise

2.4.1 Translate the following paragraph

Since statics plays an important role in both the development and application of the mechanics of materials, it is essential to have a good grasp of its fundamentals. For this reason, we will review some of the main principles of statics that will be used throughout the text.

2.4.2 Choose a proper word in the box to fill in the blanks

课文中多次出现 principle（原理）一词。在科技文献中，经常出现如概念、原理、定理、定律、法则、理论、假设等词汇，可以通过对照来学习和记忆。

| concept, principle, theorem, law, rule, theory, assumption |

① Although the (　　) of statics are time-independent, this quantity plays an important role in the study of dynamics. 尽管静力学的原理与时间无关，但这些量在动力学的研究中起重要的作用。

② Before we begin our study of engineering mechanics, it is important to understand the meaning of certain fundamental (　　) and principles. 在开始学习工程力学之前，理解特定基本概念和原理的意义是非常重要的。

③ In most cases, the actual deformations occurring in structures, machines, mechanisms, and the like are relatively small, making the rigid-body (　　) suitable for analysis. 在大多数情况下，结构、机器、机构等发生的实际变形相对较小，刚体假设适合这类问题的分析。

④ Engineering mechanics is formulated on the basis of Newton's three (　　) of motion, the validity of which is based on experimental observation. 工程力学是在牛顿三大运动定律的基础上阐述的，其正确性基于实验观察。

⑤ The main purpose of this book is to provide the student with a clear and thorough presentation of the (　　) and application of engineering mechanics. 本书的主旨是，为学生清晰、透彻地呈现工程力学理论和应用。

⑥ The magnitude of the resultant force is determined from the Pythagorean (　　). 合力的大小由毕达哥拉斯定理确定（注：毕达哥拉斯定理即勾股定理，西方普遍认为该定理是由毕达哥拉斯首先提出的）。

⑦ Subtraction can be defined as a special case of addition, so the (　　) of vector addition also apply to vector subtraction. 因为减法可以定义为一种特殊的加法，所以矢量加法法则一样可以应用于矢量减法。

Lesson 3　Introduction to Fluid Mechanics—Surface Tension

3.1　Text

It is often observed that a drop of blood forms a hump on horizontal glass; a drop of mercury forms a near-perfect sphere and can be rolled like a steel ball over a smooth surface; water droplets from rain or dew hang from branches; liquid fuel injected into an engine forms a mist of spherical droplets; water dripping from a leaky faucet falls as nearly spherical droplets; a soap bubble released into the air forms a nearly spherical shape; and water beads up into small drops on leaves [Figure 3.1(a)] A water strider has remarkable non-wetting legs that enable it to stand effortlessly and move quickly on the water [Figure 3.1(b)].

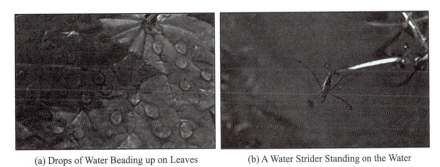

(a) Drops of Water Beading up on Leaves　　(b) A Water Strider Standing on the Water

Figure 3.1　Some Consequences of Surface Tension

In these observations, liquid droplets behave like small balloons filled with liquid, and the surface acts like a stretched elastic membrane under tension. The pulling force causing this tension acts parallel to the surface and results from the attractive forces between the liquid molecules. The magnitude of this force per unit length is called surface tension (σ_s) and is usually expressed in N/m. This effect is also called surface energy (per unit area) and is expressed in the equivalent unit of $(N \cdot m)/m^2$ or J/m^2. Here, σ_s represents the work needed to increase the surface area of the liquid by a unit amount.

[1] To visualize how surface tension arises, we present a microscopic view in Figure 3.2 by considering two liquid molecules, one at the surface and one deeper within the liquid. The attractive forces on the interior molecule by the surrounding molecules balance each other because of symmetry. However, the forces acting on the surface molecule are not

symmetric, and the attractive forces from gas molecules above are usually very small. Therefore, there is a net attractive force acting on the surface molecule, pulling it toward the liquid's interior. This force is balanced by repulsive forces from below the surface that attempt to compress the molecules. As a result, the liquid minimizes its surface area, explaining the tendency of droplets to form spherical shapes, which have the minimum surface area for a given volume.

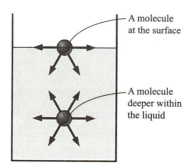

Figure 3.2　Two Liquid Molecules

You also may have observed, with amusement, that some insects can land on or even walk on water, and small steel needles can float on water. These phenomena are possible due to surface tension, which balances the weights of these objects.

[2] To understand surface tension better, consider a liquid film (such as a soap bubble film) suspended on a U-shaped wire frame with a movable side (Figure 3.3). Normally, the liquid film pulls the movable wire inward to minimize its surface area. A force F must be applied on the movable wire in the opposite direction to balance this pulling force. Both sides of the thin film are exposed to air, so the length along which surface tension acts is $2b$. Thus, a force balance on the movable wire gives $F = 2b\sigma_s$, allowing surface tension to be expressed as:

$$\sigma_s = \frac{F}{2b}$$

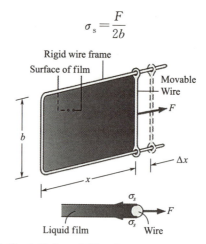

Figure 3.3　A U-shaped Wire Frame with a Movable Side

Introduction to Fluid Mechanics—Surface Tension Lesson 3

Note that for $b=0.5$m, the measured force F (in N) is equivalent to the surface tension in N/m. This kind of apparatus, with sufficient precision, can measure the surface tension of various liquids.

In the U-shaped wire frame apparatus, the movable wire is pulled to stretch the film and increase its surface area. When the wire is pulled a distance Δx, the surface area increases by $\Delta A = 2b \cdot \Delta x$, and the work W done during this process is

$$W = \text{Force} \times \text{Distance} = F \cdot \Delta x = 2b\sigma_s \cdot \Delta x = \sigma_s \cdot \Delta A$$

We assume the force remains constant over the small distance. This result also indicates the film's surface energy increases by an amount $\sigma_s \cdot \Delta A$ during stretching, which aligns with interpreting σ_s as surface energy per unit area. This is similar to a rubber band having more potential energy after stretching. For a liquid film, the work is used to move liquid molecules from the interior to the surface against the attraction forces of other molecules. Therefore, surface tension can also be defined as the work done per unit increase in the liquid's surface area.

[3] Surface tension varies significantly between substances and changes with temperature for a given substance. For example, at 20℃, the surface tension is 0.073N/m for water and 0.440N/m for mercury in atmospheric air. Mercury's high surface tension allows its droplets to form nearly spherical balls that can roll like solid balls on smooth surfaces. Generally, surface tension decreases with temperature and becomes zero at the critical point, eliminating a distinct liquid-vapor interface above this temperatures. The effect of pressure on surface tension is usually negligible.

Surface tension can be significantly altered by impurities. Surfactants, for instance, are chemicals added to liquids to reduce their surface tension. Soaps and detergents lower the surface tension of water, allowing it to penetrate small spaces between fibers for more effective cleaning. However, this also means that devices relying on surface tension, like heat pipes, can be compromised by impurities due to poor workmanship.

3.2 Words and Phrases

elastic	n. 松紧带，橡皮筋；adj. 有弹性的，灵活的
tension	n. 紧张关系，矛盾，焦虑，冲突，拉力；v. 拉紧
molecule	n. 分子
magnitude	n. 巨大，广大，重大，重要，量级，震级
equivalent	adj. 相等的，相同的；n. 等同物，对应物
symmetry	n. 对称，对称美，整齐，匀称
phenomena	n. 现象（复数）
apparatus	n. 机构，组织，器械，设备
surface energy	表面能

potential energy	势能
substance	n. 物质，材料，实质，内容，实体，（织品的）质地
critical point	临界点，紧要关头
liquid-vapor interface	液气界面
surfactant	n. 表面活性剂
detergent	n. 洗涤剂，去垢剂

3.3 Complex Sentence Analysis

[1] To visualize how surface tension arises, we present a microscopic view in Figure 3.2 by considering two liquid molecules, one at the surface and one deeper within the liquid.
① 波浪线部分是句子的主干。
② "To…"引导状语，表示目的；"by…"引导状语，表示方式、方法。
③ "one at the surface and one deeper within the liquid"进一步说明前面的 two liquid molecules。

[2] To understand surface tension better, consider a liquid film (such as a soap bubble film) suspended on a U-shaped wire frame with a movable side (Figure 3.3).
① 和上一句对照，这一句的主干部分省略了主语。同样是 To 引导目的状语。
② 两处括号，提供解释或附加说明。可以体会一下括号的展开：To understand surface tension better, consider a liquid film, such as a soap bubble film, suspended on a U-shaped wire frame with a movable side, as shown in Figure 3.3.

[3] Surface tension varies significantly between substances and changes with temperature for a given substance.
① 这是两个句子的合并。注意并列句和复合句的不同。
② 拆成两个句子：The surface tension varies significantly from substance to substance. The surface tension changes significantly with temperature for a given substance.

3.4 Exercise

3.4.1 Translate the following paragraph

Another interesting consequence of surface tension is the capillary effect, which refers to the rise or fall of a liquid in a small-diameter tube inserted into the liquid. Such narrow tubes or confined flow channels are called capillaries. The rise of kerosene through a cotton wick inserted into the reservoir of a kerosene lamp is due to this effect. The capillary effect

Introduction to Fluid Mechanics—Surface Tension　Lesson 3

also partially accounts for the rise of water to the tops of tall trees. The curved free surface of a liquid in a capillary tube is called the meniscus.

3.4.2　Choose a proper word in the box to fill in the blanks

课文中大量出现单词 tension，意为"张力"，它由词根 "tens" 变化而来。"tens" 的常见构词如下：

tense → tension, tensive, tensile, tensible

intense → intension, intensive, intensity, intensify

extense → extension, extensive

> tensile, intensive, intensity, tension, extension, extensive

① The surface of the liquid acts like a stretched elastic membrane under (　　). 在张力的作用下，液体表面表现得像一张被拉紧的弹性薄膜。

② In an oscillating incompressible flow field the force per unit mass acting on a fluid particle is obtained from Newton's second law in (　　) form. 在振荡的不可压缩流场中，作用在流体质点上的单位质量力可以通过牛顿第二定律的密度形式得到。

③ The eddy motion loses its (　　) close to the wall and diminishes at the wall because of the no-slip condition. 由于无滑移条件，涡流运动会在靠近壁面处失去强度，并在壁面处降低。

④ Taking the average friction factor to be 0.02, calculate the mass flow rate, and the inlet velocity, for various (　　) lengths. 取平均摩擦系数为 0.02，计算不同延长长度下的质量流量和入口速度。

⑤ This chapter makes (　　) use of force balances for bodies in static equilibrium. 本章广泛利用力平衡方法分析静态平衡的物体。

⑥ Remember that forces that are found to pull on the joint are (　　) forces, and those that push on the joint are compressive forces. 请记住，拉动关节的力是拉伸力，而推动关节的力是压缩力。

Lesson 4　Machine Parts（Ⅰ）

4.1　Text

1. Gears

【拓展视频】

[1]Gears are direct contact bodies, operating in pairs, that transmit motion and force from one rotating shaft to another, or from a shaft to a slide (rack), by means of successively engaging projections called teeth.

The contacting surfaces of gear teeth must be aligned so that the drive is positive; i. e., the load transmitted must not depend on frictional contact. As shown in the treatment of direct contact bodies, this requires that the common normal to the surfaces does not pass through the pivotal axis of either the driver or the follower.

Known as direct contact bodies, cycloidal and involute profiles provide both a positive drive and a uniform velocity ratio; i. e., conjugate action.

The smaller gear in a pair is called the pinion, and the larger is the gear. When the pinion is on the driving shaft, the pair acts as a speed reducer; when the gear drives, the pair is a speed increaser. Gears are more frequently used to reduce speed than to increase it.

If a gear having N teeth rotates at n revolutions per minute, the product $N \times n$ has the dimension "teeth per minute". This product must be the same for both members of a mating pair if each tooth acquires a partner from the mating gear as it passes through the region of tooth engagement.

For conjugate gears of all types, the gear ratio and the speed ratio are both given by the ratio of the number of teeth on the gear to the number of teeth on the pinion. If a gear has 100 teeth and a mating pinion has 20, the ratio is $100/20=5$. Thus, the pinion rotates five times as fast as the gear, regardless of the speed of the gear. Their point of tangency is called the pitch point, and it is the only point at which the tooth profiles have pure rolling contact since it lies on the line of centers. Gears on nonparallel and non-intersecting shafts also have pitch circles, but the rolling-pitch-circle concept is not valid.

Gear types are largely determined by the disposition of the shafts; this means that if a specific disposition of the shafts is required, the type of gear will more or less be fixed. If a required speed change demands a certain type, the shaft positions will also be fixed.

A gear with tooth elements that are straight and parallel to its axis is known as a spur gear. A spur gear can be used to connect parallel shafts only.

[2] If an involute spur pinion were made of rubber and twisted uniformly so that the ends rotated about the axis relative to one another, the elements of the teeth, initially straight and parallel to the axis, would become helices. The pinion would then effectively become a helical gear.

To achieve line contact and improve the load-carrying capacity of crossed axis helical gears, the gear can be made to curve partially around the pinion, in somewhat the same way that a nut envelops a screw. The result would be a cylindrical worm and gear. Worms are also made in the shape of an hourglass, instead of cylindrical, so that they partially envelop the gear. This results in a further increase in load-carrying capacity.

Worm gears provide the simplest means of obtaining large ratios in a single pair. They are usually less efficient than parallel-shaft gears, however, because of additional sliding movement along the teeth.

2. V-belt

Rayon and rubber V-belts are widely used for power transmission. These belts are made in two series: the standard V-belt and the high capacity V-belt. They can be used with short center distances and are made endless to avoid difficulties with splicing devices.

【拓展视频】

First, the cost is low, and power output can be increased by operating several belts side by side. All belts in the drive should stretch at the same rate to keep the load equally divided among them. When one belt breaks, the entire group usually needs replacement. The drive can be inclined at any angle with the tight side either on the top or bottom. Since belts can operate on relatively small pulleys, large reductions in speed in a single drive are possible.

【拓展视频】

Second, the included angle for the belt groove is usually between 34° and 38°. The wedging action of the belt in the groove significantly increases the tractive force developed by the belt.

Third, pulley may be made of cast iron, sheet steel, or die-cast metal. [3] Sufficient clearance must be provided at the bottom of the groove to prevent the belt from bottoming as it becomes narrower from wear. Sometimes the larger pulley is left ungrooved when it is possible to develop the required tractive force by running on the inner surface of the belt. This eliminates the cost of cutting the grooves. Some pulleys on the market allow for adjustment in the width of the groove. This varies the effective pitch diameter of the pulley, enabling moderate changes in the speed ratio.

3. Chain Drives

The first chain-driven bicycle appeared in 1874, and chains were used for driving the rear wheels on early automobiles. [4] Today, as the result of modern design and production methods, chain drives far superior to their prototypes are available, and they have greatly contributed to the

【拓展视频】

【拓展视频】

development of efficient agricultural machinery, well-drilling equipment, and mining and construction machinery. Since about 1930, chain drives have become increasingly popular, especially for power saws, motorcycle, escalators, etc.

There are at least six types of power-transmission chains; three of these the roller chain, the inverted tooth or silent chain, and the bead chain will be covered as follows:

(1) Roller chain.

The essential elements in a roller-chain drive are a chain with side plates, pins, bushings (sleeves), rollers, and two or more sprocket wheels with teeth resembling gear teeth. Roller chains are assembled from pin links and roller links. A pin link consists of two side plates connected by two pins inserted into holes in the side plates. The pins fit tightly into the holes, forming what is known as a press fit. A roller link consists of two side plates connected by two press-fitted bushings, on which two hardened steel rollers are free to rotate. When assembled, the pins are a free fit in the bushings and rotate slightly relative to the bushings as the chain engages and disengages with a sprocket.

Standard roller chains are available in a single strand or multiple strands. In the latter type, two or more chains are joined by common pins that keep the rollers in the separate strands properly aligned. The speed ratio for a single drive should be limited to about 10∶1; the preferred shaft center distance is 30 to 35 times the distance between the rollers, and chain speeds greater than about 800m/min are not recommended. Roller chains are particularly well suited for driving several parallel shafts from a single shaft without slip.

(2) Inverted tooth or silent chain.

An inverted tooth or silent chain is essentially an assemblage of gear racks, each with two teeth, pivotally connected to form a closed chain with the teeth on the inside, meshing with conjugate teeth on the sprocket wheels. The links are pin-connected flat steel plates, usually with straight-sided teeth having an included angle of 60°, as many links are connected side by side to transmit the power. Compared with roller-chain drives, silent-chain drives are quieter, operate successfully at higher speeds, and can transmit more loads for the same width. Some automobiles use silent-chain camshaft drives.

(3) Bead chain.

Bead chains provide an inexpensive and versatile means for connecting parallel or nonparallel shafts when the speed and power transmitted are low. The sprocket wheels contain hemispherical or conical recesses into which the beads fit. The chains resemble key chains and are available in plain carbon and stainless steel, as well as in the form of solid plastic beads molded on a cord. Bead chains are used in computers, air conditioners, television tuners, and Venetian blinds. The sprockets may be made of steel, die-cast zinc or aluminum, or molded nylon.

4.2 Words and Phrases

gear	n. 齿轮
slide	n. 滑块
rack	n. 齿条
projection	n. 凸出，凸起部分
cycloidal	adj. 摆线的
cycloidal profile	摆线轮廓
involute	adj. 渐开线的
involute profile	渐开线轮廓
conjugate	adj. 共轭的
pinion	n. 小齿轮
dimension	n. 量纲，尺寸
mate	v. 啮合，配对
engagement	n. 啮合
tangency	n. 接触
pitch	n. 齿节
intersect	v. 相交，交叉
disposition	n. 排列，配置
helical gear	螺旋齿轮，斜齿轮
spur gear	正齿轮，直齿轮
worm	n. 蜗轮，蜗杆
bevel gear	n. 伞形齿轮，锥齿轮
hourglass	n. 沙漏
V-belt	n. V形带
pulley	n.（皮带）轮
groove	n. 沟，槽
tractive	adj. 牵引的
clearance	n. 间隙
chain drive	链传动
prototype	n. 模型，原型机
saw	n. 锯
escalator	n. 自动扶梯
roller chain	套筒滚子链条，滚子链
bead chain	滚珠链条
bushing	n. 套筒，衬套
sprocket	n. 链轮

strand	*n.* 排，列
venetian blind	威尼斯百叶窗，软百叶窗
die-cast	*n.* 压铸品；*adj.* 压铸的

4.3 Complex Sentence Analysis

[1] Gears are direct contact bodies, operating in pairs, that transmit motion and force from one rotating shaft to another, or from a shaft to a slide (rack), by means of successively engaging projections called teeth.

① operating in pairs 为分词短语，修饰 Gears。
② that 引导的从句修饰 Gears。
③ by means of：借助，通过。

[2] If an involute spur pinion were made of rubber and twisted uniformly so that the ends rotated about the axis relative to one another, the elements of the teeth, initially straight and parallel to the axis, would become helices.

① were made of：由……制成。
② so that 引导结果状语从句。
③ parallel to…：平行于……

[3] Sufficient clearance must be provided at the bottom of the groove to prevent the belt from bottoming as it becomes narrower from wear.

① at the bottom of…：在……的底部。
② prevent…from wear：防止磨损。

[4] Today, as the result of modern design and production methods, chain drives far superior to their prototypes are available, and they have greatly contributed to the development of efficient agricultural machinery, well-drilling equipment, and mining and construction machinery.

① superior to…：优于……
② and 连接两个并列分句。

4.4 Exercise

4.4.1 Translate the following paragraphs

A gear with tooth elements that are straight and parallel to its axis is known as a spur gear. A spur gear can only be used to connect parallel shafts. Parallel shafts, however, can also be connected by gears of another type, and a spur gear can be paired with a gear of a different type.

Helical gears have certain advantages; for example, when connecting parallel shafts, they have a higher load-carrying capacity than spur gears with the same number of teeth cut with the same cutter. Helical gears can also be used to connect nonparallel, non-intersecting shafts at any angle to one another, with 90° being the most common angle for such applications.

Worm gears provide the simplest means of obtaining large ratios in a single pair. They are usually less efficient than parallel shaft gears; however, because of additional sliding movement along the teeth. Due to their similarities, the efficiency of a worm gear depends on the same factors as the efficiency of a screw.

4.4.2 Write the name of the following parts and brief description of the roller chain on the structure (Figure 4.1) and assembly relations

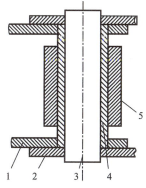

Figure 4.1 Structure

Lesson 5   Machine Parts (Ⅱ)

5.1   Text

1. Fastener

[1] Fasteners are devices which permit one part to be joined to a second part, and they are involved in almost all designs.

There are three main classifications of fasteners, described as follows:

(1) Removable fastener. This type of fastener permits the parts to be readily disconnected without damaging the fastener. An example is the ordinary nut-and-bolt fastener.

(2) Semi-permanent fastener. For this type, the parts can be disconnected, but some damages usually occur to the fastener. One example is a cotter pin.

(3) Permanent fastener. This type of fastener is intended for parts that will never be disassembled. Examples include riveted joints and welded joints.

The importance of fasteners becomes evident when considering any complex products. In the case of an automobile, there are literally thousands of parts fastened together to produce the complete product. The failure or loosening of a single fastener could result in a simple nuisance such as a door rattle or in a serious situation such as a wheel coming off. Such possibilities must be taken into account when selecting a type of fastener for a specific application.

Nuts, bolts, and screws are undoubtedly the most common means of joining materials. Since they are so widely used, it is essential that these fasteners achieve the maximum effectiveness at the lowest possible cost.

An ordinary nut loosens when the forces of vibration overcome those of friction. In a nut and lock washer combination, the lock washer provides an independent locking feature, preventing the nut from loosening. The lock washer is useful only when the bolt might loosen because of a relative change between the length of the bolt and the parts it assembles. [2] This change in the length of the bolt can be caused by a number of factors: creep in the bolt, loss of resilience, differences in thermal expansion between the bolt and the bolted members, or wear. In these static cases, the expanding lock washer holds the nut under axial load and keeps the assembly tight. When relative changes are caused by vibration forces, the lock washer is not nearly as effective.

Rivets are permanent fasteners. They rely on deformation of their structure for their holding action. Rivets are usually stronger than thread-type fasteners and are more economical on a first-cost basis. Rivets are driven either hot or cold, depending on the mechanical properties of the rivet material. Aluminum rivets, for instance, are cold-driven, as cold working improves the strength of aluminum. Most large rivets, however, are hot-driven.

2. Shaft

Virtually almost all machines contain shafts. The most common shape for shafts is cylinder, and the cross section can be either solid or hollow, with hollow shafts offering potential weight savings.

【拓展视频】

Shafts are mounted in bearings and transmit power through devices such as gears, pulleys, cams, and clutches. These devices introduce forces which attempt to bend the shaft; hence, the shaft must be rigid enough to prevent overloading of the supporting bearings. [3] In general, the bending deflection of a shaft should not exceed 0.01in (about 0.00025m) per foot (about 12in or 0.30m) of length between bearing supports.

For diameters less than 3in (about 0.08m), the usual shaft material is cold-rolled steel containing about 0.4% carbon. Shafts are either cold-rolled or forged in sizes from 3 to 5in (about from 0.08 to 0.13m). For sizes above 5in (about 0.13m), shafts are forged and machined to size. Plastic shafts are widely used for light load applications. One advantage of using plastic is safety in electrical applications, as plastic is a poor conductor of electricity.

Another important aspect of shaft design is the method of directly connecting one shaft to another. This is accomplished using devices such as rigid and flexible couplings.

3. Bearing

A bearing can be defined as a component member specifically designed to support moving machine parts. The most common application is the support of a rotating shaft that transmits power from one location to another. Since there is always relative motion between a bearing and its mating surface, friction is involved. In many instances, such as the design of pulleys, brakes, and clutches, friction is desirable. However, in bearings, reducing friction is a primary consideration because it results in power loss, heat generation, and increased wear of mating surfaces.

The concern of a machine designer with ball bearings and roller bearings are life in relation to load, stiffness or deflections under load, friction, wear, and noise. For moderate loads and speeds, the correct selection of a standard bearing based on load rating will usually ensure satisfactory performance. The deflection of the bearing elements becomes important under high loads, though this is usually less significant than deflection of the shafts or other components associated with the bearing. [4] Where speeds are high, special cooling arrangements which may increase frictional drag become necessary. Wear is primarily associated with the introduction of contaminants, so sealing arrangements must be chosen with regard to the environment's hostility.

Although the responsibility for the basic design of ball bearings and roller bearings rests with the manufacturer, the machine designer must understand the duty to be performed by the bearing. They must be concerned not only with bearing selection but also with the conditions for correct installation.

The fit of the bearing races onto the shaft or housings is critical because of their combined effect on the internal clearance of the bearing and the desired degree of interference fit. Inadequate interference can cause serious trouble from fretting corrosion. The inner race is frequently located axially by abutting against a shoulder. A radius at this point is essential to avoid stress concentration, and ball races are provided with a radius or chamfer to accommodate this.

A journal bearing, in its simplest form, is a cylindrical bushing made of suitable material with properly machined inside and outside diameters. The journal is usually the part of a shaft or pin that rotates inside the bearing, as shown in Figure 5.1.

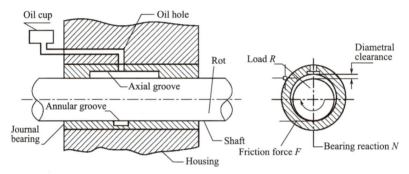

Figure 5.1　Journal Bearing

Journal bearings operate with sliding contact, and to reduce the issues associated with sliding friction, a lubricant is used along with compatible mating materials. When selecting the lubricant and mating materials, bearing pressures, temperatures, and rubbing velocities must be considered. The principle function of the lubricant in sliding contact bearings is to prevent physical contact between the rubbing surfaces. Thus, maintaining an oil film under varying loads, speeds, and temperature is the prime consideration in sliding contact bearings.

5.2　Words and Phrases

device	*n.* 设备，装置
fastener	*n.* 紧固件，紧固零件
classification	*n.* 分类，类别
removable	*adj.* 可移动的，可拆的
semipermanent	*adj.* 半永久性的

Machine Parts (II) Lesson 5

cotter pin	开口销，开尾销
rivet	n. 铆钉；v. 铆，铆接
weld	v. 焊接，熔接
nuisance	n. 障碍，损害
rattle	v. 发出咯咯声；n. 硬物质的撞击声
nut	n. 螺帽
bolt	n. 螺栓；v. 用螺栓连接
screw	n. 螺钉，螺旋丝杆
lock washer	锁紧垫圈
resilience	n. 弹性
aluminum	n. 铝
shaft	n. 轴
bearing	n. 轴承
cam	n. 凸轮
clutch	v. 抓住；n. 离合器
cold-roll	v. 冷轧；n. 冷轧机
forge	v. & n. 锻造，打制
flexible	adj. 柔软的
friction	n. 摩擦
break	v. 破坏，折断，损坏
wear	v. & n. 磨损，耗损
arrangement	n. 布置，排列
contaminant	n. 杂质，污染物质
sealing arrangement	密封装置
hostility	n. 敌意，恶劣
appreciation	n. 评价，欣赏
interference	n. 干涉，过盈
fretting	n. 微振磨损
corrosion	n. 腐蚀
abut	v. 邻接，倚靠
stress concentration	应力集中
shoulder	n. 轴肩
chamfer	v. & n. 倒角，倒圆，开槽
journal bearing	滑动轴承
cylindrical	adj. 圆筒状的，柱状的
lubricant	n. 润滑剂
compatible	adj. 兼容的，和谐的，一致的

5.3　Complex Sentence Analysis

［1］Fasteners are devices which permit one part to be joined to a second part, and they are involved in almost all designs.

① which 引导定语从句，修饰 devices。

② to be joined to a second part：第一个不定式 to 表示目的；be joined to 是介词词组，to 表示"到"的意思。

③ are involved in：涉及，包括。

［2］This change in the length of the bolt can be caused by a number of factors：creep in the bolt, loss of resilience, differences in thermal expansion between the bolt and the bolted members, or wear.

loss of resilience：弹性丧失。

［3］In general, the bending deflection of a shaft should not exceed 0.01in (about 0.00025m) per foot (about 0.25m) of length between bearing supports.

① 0.01in (about 0.00025m) per foot (about 0.25m)：每英尺长度上为 0.01 英寸。1 英尺约为 0.25m，1 英寸约为 0.00025m。

② between bearing supports：between 表示两者之间，可译为"在两轴承支承之间"。

［4］Where speeds are high, special cooling arrangements which may increase frictional drag become necessary.

which may increase frictional drag 修饰主语 cooling arrangements。

5.4　Exercise

Translate the following paragraphs

A bearing is a connector that allows the connected part to either rotate or translate relative to one another, while preventing them from separating in the direction of applied loads. Often, one of the parts is fixed, and the bearing supports the moving element.

Sliding bearings are the simplest to construct and, given the numerous pin-jointed devices and structures in use, are probably the most commonly used.

The essential components of a ball bearing are the inner and outer rings, the balls, and the separator. The inner ring is mounted on a shaft and has a groove for the balls to roll in. The outer ring is usually stationary and also contains a groove to guide and support the balls. The separator prevents contact between the balls, reducing friction, wear, and noise in areas where sliding might occur.

Lesson 6 Mechanisms

6.1 Text

A mechanism is a combination of two or more connected parts designed to achieve a specific motion. These are components of machinery. The connections between two parts with relative motion are called motion pairs.
【拓展视频】
Motion pairs that contact with surfaces are called lower pairs, while those that contact with points or lines are called high pairs. [1] The specific motion properties of a mechanism depend on the relative size of the parts, the nature of the motion pairs, and their arrangement. A part used to support motion in the mechanism is called the machine frame and acts as the reference point for studying the motion system. The part with independent motion is called the motivity member, while other parts, except the machine frame and motivity member that are compelled to move are called the driven member. The number of independent parameters needed to describe the mechanism's motion is called the degrees of freedom. To achieve specific relative motion, the number of motivity members must equal the degrees of freedom.

Mechanisms can be categorized into planar, spherical, and spatial types. They share common characteristics; [2] the criterion which distinguishes the types, however, is to be found in the motion of their links.

A planar mechanism is one where all particles describe plane curves in space, and these curves lie in parallel planes; i.e., the loci of all points are plane curves parallel to a single common plane. [3] This characteristic makes it possible to represent any chosen point of a planar mechanism in its true size and shape on a single drawing or figure. The motion transformation of such a mechanism is called coplanar. Examples include the plane four-bar linkage, plate cam and driven parts, and slider-crank mechanism. Most mechanisms in use today are planar.

A cam is a machine member that drives a follower through a specified motion. Proper cam design allows for the desired motion of a machine member. Cams are widely used in machinery, including internal combustion engines, machine tools, compressors, and computers. In general, a cam can be designed in two ways:

(1) The profile of a cam is designed to give a desired motion to the follower.

(2) A suitable profile is chosen to ensure satisfactory performance by the follower.

A rotary cam is a machine part which changes cylindrical motion into straight-line

【拓展视频】

motion. Its purpose is to transmit various kinds of motion to other parts of a machine.

Practically every cam must be designed and manufactured to fit special requirements. Although each cam may differ, they operate similarly. As a cam rotates, it moves a connected part called a follower either right or left, up or down, and in or out. The follower is usually connected to other parts of the machine to achieve the desired action. If the follower loses contact with the cam, it will fail to work.

Cams are classified by their shape into four types: plate (disc) cam, translation cam, cylindrical cam, and face cam (Figure 6.1).

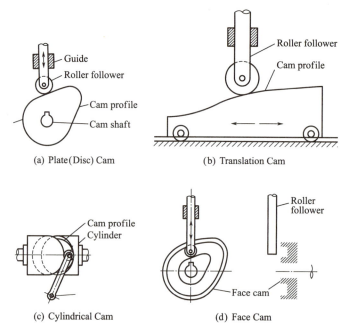

Figure 6.1　Types of Cam

Planar mechanisms utilizing only lower pairs are called planar linkage mechanisms, which may include only revolute and prismatic pairs. Although a planar pair might theoretically be included, it would impose no constraint and be equivalent to an opening in the kinematics chain. Planar motion also requires that the axes of all prismatic pairs and revolute axes be normal to the plane of motion.

6.2　Words and Phrases

mechanism　　　　　　　　n. 机构
motion pairs　　　　　　　运动副
activity connection　　　　活动连接件

Mechanisms Lesson 6

lower pair	低副
higher pair	高副
disposition	*n.* 配置，排列
machine frame	机座，机架
coordinate	*n.* 坐标
driving member	原动件
parameter	*n.* 参变量，参数
driven member	从动件
degrees of freedom	*n.* 自由度
categorize	*v.* 分类
category	*n.* 种类，逻辑范畴
planar	*adj.* 平面的
spherical	*adj.* 球的，球面的
spatial	*adj.* 空间的
loci	*n.* ［locus 的复数形式］点的轨迹
constraint	*n.* 约束
prismatic pair	移动副

6.3 Complex Sentence Analysis

［1］ The specific motion properties of a mechanism depend on the relative size of the parts, the nature of the motion pairs, and their arrangement.

　　specific motion properties：运动特性。

［2］ the criterion which distinguishes the types, however, is to be found in the motion of their links.

① which 引导定语从句，修饰 criterion。

② to be found 为不定式被动语态。

③ links：连杆装置。

［3］ This characteristic makes it possible to represent any chosen point of a planar mechanism in its true size and shape on a single drawing or figure

① makes it possible：使其成为可能。

② represent：描绘，展现。

③ planar mechanism：平面机构。

④ in its true size and shape：在它真实的尺寸和形状方面。

6.4　Exercise

Translate the lineate phrases and fill in the brackets

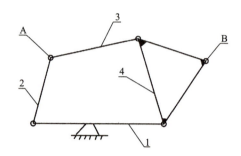

该机构为平面四连杆机构（　　　），
1 为机架（　　　），
2 为原动件（　　　），
3 和 4 为从动件（　　　），
A 为回转副（　　　），属于低副（　　　），
B 为固定连接（　　　）。
构件（　　　）4 中的焊接符号（　　　）表示 4 为一个构件。
该机构的自由度（　　　）只有等于原动件数才能实现确定的运动（　　　）。

Lesson 7 Fluid and Hydraulic System

7.1 Text

[1] The history of hydraulic power is a long one, dating back to humanity's prehistoric efforts to harness natural energy. The primaly sources available were water and wind—two free and moving streams.

The watermill, the first hydraulic motor, was an early invention. One is depicted in a mosaic at the Great Palace in Byzantium from the early 5th century, built by the Romans. However, the first record of a watermill goes back even further, to around 100 BC, and its origins may be even earlier. 【拓展视频】 Grain domestication began around 5000 years ago, and some enterprising farmer, or perhaps a farmer's wife, likely became tired of grinding grain by hand and sought a better method.

A fluid is a substance that can flow; its particles can continuously change their positions relative to each other, offering no lasting resistance to the displacement of one layer over another. This means that if the fluid is at rest, no shear force (a force tangential to the surface on which it acts) can exist in it.

Fluids can be classified as Newtonian or non-Newtonian. [2] In a Newtonian fluid, there is a linear relationship between applied shear stress and the resulting rate of angular deformation. In a non-Newtonian fluid, this relationship is nonlinear.

Fluids flow can be classified in many ways, such as steady or unsteady, rotational or irrotational, compressible or incompressible, and viscous or non-viscous.

[3] All hydraulic systems depend on Pascal's law, named after Blaise Pascal, who discovered the law. This law states that a pressurized fluid within a closed container, such as a cylinder or a pipe, exerts equal force on all surfaces of the container.

In hydraulic systems, Pascal's law defines the basis for the results obtained. A pump moves the liquid through the system, with its intake connected to a liquid source, usually called the tank or reservoir. Atmospheric pressure forces the liquid into the pump. When operating, the pump forces the liquid from the tank into the discharge pipe at suitable pressure.

The flow of the pressurized liquid is controlled by valves. Three control functions are used in most hydraulic systems: control of liquid pressure, control of liquid flow rate, and control of the direction of liquid flow.

Hydraulic drives are preferred over mechanical systems when:

【拓展视频】

(1) power needs to be transmitted over long distances unsuitable for chains or belts.

(2) high torque at low speed is required.

(3) a compact unit is needed.

(4) smooth, vibration-free transmission is required.

(5) easy control of speed and direction is necessary.

(6) output speed needs to be varied seamlessly.

Figure 7.1 shows speed control of hydraulic motors.[4] Electrically driven oil pressure pumps establish an oil flow for energy transmission, which is fed to hydraulic motors or cylinders, converting it into mechanical energy. The oil flow is controlled by valves, producing linear or rotary mechanical motion. The kinetic energy of the oil flow is comparatively low; therefore, the term hydrostatic driver is sometimes used. There is little difference between hydraulic motors and pumps; any pump may be used as a motor. The oil flow quantity can be varied using regulating valves or variable-delivery pumps.

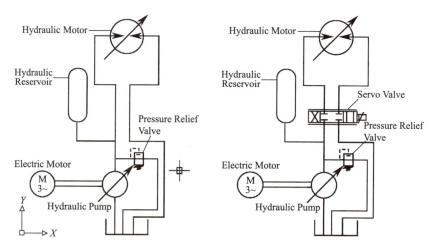

Figure 7.1 Speed Control of Hydraulic Motors

【拓展视频】

The application of hydraulic power to the operation of machine tools is not new, although its widespread adoption is comparatively recent. The development of modern self-contained pump units stimulated the growth of this form of machine tool operation.

Hydraulic machine tool drives offer many advantages, such as infinitely variable speed control over wide ranges and the ability to change the direction of drive as easily as varying the speed. Many complex mechanical linkages can be simplified or eliminated by hydraulics.

The flexibility and resilience of hydraulic power offer significance. In addition to smooth operation, there is often an improvement advantages, in the surface finish of the work. The tool can make heavier cuts without detriment and lasts considerably longer without requiring regrinding.

7.2　Words and Phrases

hydraulic system	液压系统
displacement	$n.$ 位移，转移，置换
layer	$n.$ 层，层次
tangential	$adj.$ 切线的，切向的
Newtonian	$adj.$ 牛顿的，牛顿学说的
nonlinear	$adj.$ 非线性的，非直线的
rotational	$adj.$ 旋转的，转动的，循环的
compressible	$adj.$ 可压缩的，可压榨的
Pascal's law	帕斯卡定律
intake	$n.$ 入口，进口，进入量
tank	$n.$ 油箱，水箱，池塘
reservoir	$n.$ 蓄水池，水箱，蓄能器
atmospheric	$adj.$ 大气的，空气的
discharge	$n.$ 卸货，出料，流出；$vi.$ 卸下，放出
pressurize	$v.$ 增压，给……加压
prehistoric	$adj.$ 史前的，很久以前的
harness	$v.$ 利用（风等）作动力，治理，控制
watermill	$n.$ 水车，水磨
mosaic	$n.$ 镶嵌细工，马赛克
domestication	$n.$ 家养，驯养
preference	$n.$ 优先选择
compact	$adj.$ 紧凑的，紧密的，简洁的
diagrammatic	$adj.$ 图表的，概略的
oil pressure pump	油泵
hydraulic motor	液压电动机
hydraulic cylinder	油缸
kinetic energy	动能
hydrostatic driver	静压传动
variable-delivery pump	变量泵
by no means	决不……
self-contained	独立的，配套的，整体的
stimulate	$v.$ 促进，激励

hydraulics	*n.* 水力学，液压系统
resilience	*n.* 跳回，恢复力，回弹
virtue	*n.* 优点，效力，功能
detriment	*n.* 损害，不利
regrind	*v.* 重磨

7.3　Complex Sentence Analysis

［1］ The history of hydraulic power is a long one, dating back to humanity's prehistoric efforts to harness natural energy.
① dating back to…：追溯到……
② prehistoric efforts：很久以前的努力。
［2］ In a Newtonian fluid, there is a linear relation ship between applied shear stress and the resulting rate of angular deformation.
① between…and…：在……和……之间。
② applied shear stress：作用的剪切应力。
③ the resulting rate：总合率（量）。
［3］ All hydraulic systems depend on Pascal's law, named after Blaise Pascal, who discovered the law.
① depend on：依赖，取决于，遵循。
② named after…：根据……命名的。
③ who discovered the law 引导非限定性定语从句，修饰 Braise Pascal。
［4］ Electrically driven oil pressure pumps establish an oil flow for energy transmission, which is fed to hydraulic motors or hydraulic cylinders, converting it into mechanical energy.
① which is fed to hydraulic motors or hydraulic cylinders 中的 which 是指 oil flow。
② converting…into…：将……转换为……

7.4　Exercise

Translate the following paragraphs

Compressors are used in petrochemical plants to increase the static pressure of air and process gases to levels required to overcome pipe friction, facilitate a certain reaction at the point of delivery, or impart desired thermodynamic properties to the compressed medium. These compressors come in a variety of sizes, types, and models, each designed to meet specific needs and likely representing the optimum configuration for a given set of requirements.

The selection of compressor types must be preceded by a comparison between service requirements and compressor capabilities. This initial comparison generally leads to a review of the economies of space, installation cost, operating cost, and maintenance requirements of competing types. When the superiority of one compressor type or model over another is not obvious, a more detailed analysis may be warranted.

Lesson 8　Engineering Graphic

8.1　Text

[1] "Graphics" comes from the Greek word "grapho", meaning "drafting" or "drawing". Drawing is the primary medium for developing and communicating technical ideas. Engineering drawings provide an exact and complete description of objects, including dimensions, tolerances, and other manufacturing information. Thus, engineering drawing is often referred to as the common language of engineering, and every engineer must master this language. The main tasks of engineering graphics indude:

(1) Learning projection techniques.

(2) Developing drawing creation and interpretation skills.

(3) Cultivating spatial analysis and visualization abilities.

As a common language of engineering, drawing is used to direct production and facilitate technical interchange. Therefore, it is necessary to standardize drafting practices, such as layout and dimension.

1. Formation of Three-projection Views

To determine the position of a point, more projection planes are added. Usually, three perpendicular planes are used in orthographic projection. They are horizontal projection plane, frontal projection plane, and profile projection plane, denoted by H, V, and W, respectively.

Three projection planes divide space into eight quadrants, as shown in Figure 8.1. [2] According to GB/T 14692—2008 "Technical drawings-Projection methods", the first-angle projection is used to make engineering drawings while in some other countries, such as in the USA and Canada, the third-angle projection is used. In this paper we focus on the third-angle projection.

In the first-angle projection, an object is placed in the quadrant Ⅰ, and observer always looks through the object towards the projection plane. In the third-angle projection, the object is placed in the quadrant Ⅲ, and observer always looks through the projection plane towards the object. The projection plane is assumed to be transparent, forming views.

The first-angle and the third-angle projection are shown in Figure 8.2.

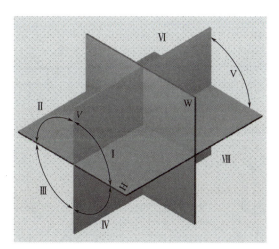

Figure 8.1 Eight Quadrants

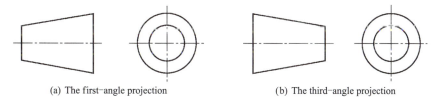

(a) The first-angle projection   (b) The third-angle projection

Figure 8.2 The Projection Identification Symbol

2. Composite Objects

Projection rules are: Front and top views are aligned vertically to show the width of the object; right and left views are aligned horizontally to show the height of the object; top and right views share the same depth of the object.

Methods to draw three views are as follows:

(1) Analyzing-shape method. Any composite object can be broken into a combination of some primary geometric shapes, which can be classified into superposition and cutting style.

(2) Select the projection. The front view is crucial, so select an adequate projection direction for it.

(3) Drawing steps: Locate axis lines, center lines of symmetry, and base lines; draw the base with an H pencil; check the drawing and darken lines.

Methods to read composite views are as follows:

(1) Break the object into basic shapes.

(2) Simultaneously interpret at least two views.

(3) Understand the meanings of lines and areas.

Methods to read views are as follows:

(1) Analyzing shape method. Break the object into geometric solids.

(2) Analyzing lines and planes method. Break the object into surfaces and lines. Three views of a composite object is shown in Figure 8.3.

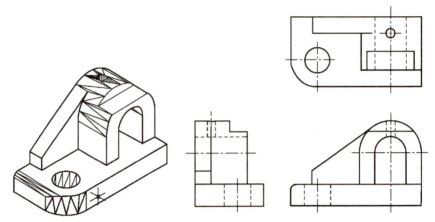

Figure 8.3   Three Views of A Composite Object

3. Detail Drawings

Detail drawings indicate the construction, size, and technical requirements of a part, describing its shape and providing necessary manufacturing information.

Contents of a detail drawing indude:

(1) Sufficient views for a complete shape description.

(2) All dimensions for manufacturing.

(3) Technical requirements, including form tolerances, position tolerances, surface roughness, material specification, heat treatments, etc.

Principles of selecting views are as follows:

(1) To meet the requirements of making a clear and complete shape description of the object, determine required views and the best way to position the part.

(2) Select main view, based on characteristic shape, functioning position, and machining position.

(3) Select other views, limit to necessary views, avoid hidden lines and repetition.

4. Assembly Drawings

An assembly drawing shows machine parts in their working positions.

Contents of an assembly drawing indude:

(1) Views showing positional relationships and operations.

(2) Necessary dimensions for critical part positions and site positioning.

(3) Technical requirements for assembling, checking, and maintenance.

(4) Item numbers, item list, and title block.

Conventions in assembly drawings indude:

(1) General conventions: No gap between contact surfaces; gaps shown between non-contact surfaces; different section line directions for adjacent pares; solid parts cut along

their axis shown without section lines, such as shafts, axles, rods, handles, pins, and keys; screws, bolts, nuts, and their washers kept in their shapes.

(2)[3] Special conventions: Representations of cuts along joint faces, separate parts, phantom lines, exaggerated representations, and simplified representations.

8.2 Words and Phrases

graphics	n. 制图，图学
drafting	n. 草图，制图
drawing	n. 绘图，制图，图样
projection	n. 投影
dimension	n. 尺寸；v. 给……标注尺寸
spatial analysis	空间分析
spatial visualization	空间想象
horizontal projection	水平投影
frontal projection	正投影
profile projection	侧投影
quadrant	n. 象限
center lines of symmetry	对称中心线
composite object	组合体
detail drawing	零件图
assembly drawing	装配图
phantom line	假想线

8.3 Complex Sentence Analysis

[1] "Graphics" comes from the Greek word "grapho", meaning "drafting" or "drawing". Drawing is the primary medium for developing and communicating technical ideas.

① meaning "drafting" or "drawing" 为动名词，作伴随状语，"drafting" or "drawing" 为动名词，作 meaning 的宾语。

② Drawing is the primary medium for developing and communicating technical ideas 中的 drawing 为动名词，作主语。

[2] According to the Chinese National Standard of Technical Drawings, the first-angle projection is used to make engineering drawings while in some other countries, such as in the USA and Canada, the third-angle projection is used. according to 表示按照（某惯例、某条款、某规定、某法律法规、某情况等）。

[3] Special conventions: Representations of cuts along joint faces, separate parts, phantom lines, exaggerated and simplified representations.

① special conventions：特殊画法。特殊画法与通用画法相比，应用较少，特殊情况下才需要用到。例如，装配图主要是表达各零件之间的装配位置关系的，通常无须单独表达某个具体的零件结构；但如果这个零件上的孔槽较多，为了更好的表达与之装配的其他零件的位置，就需要采用特殊画法。

8.4 Exercise

8.4.1 Translate the following paragraph

Similar to an offset in that the cutting-plane line staggers, it differs because the cutting-plane line is offset at an angle other than 90°. When the section is taken, the sectional view is drawn as if the cutting-plane is rotated to be perpendicular to the line of sight. This can cause the right-side sectional view to sometimes appear elongated, depending on the shape.

8.4.2 Draw the front view to an aligned section.

Figure 8.4 shows the topic, and Figure 8.5 shows the answer.

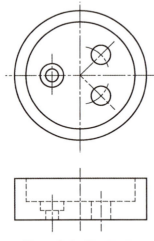

Figure 8.4　The Topic

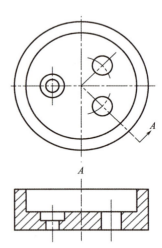

Figure 8.5　The Answer

Lesson 9　Introduction to CAD/CAM/CAPP

9.1　Text

Throughout the history of our industrial society, many inventions have been patented, and whole-new technologies have evolved. [1]Perhaps the single development that has impacted manufacturing more quickly and significantly than any previous technology is the digital computer. Computers are increasingly used for both the design and detailing of engineering components in the drawing office.

[2]Computer aided design (CAD) is defined as the application of computers and graphics software to aid or enhance the product design from conceptualization to documentation. CAD is the most commonly associated with the use of an interactive computer graphics system, referred to as a CAD system. CAD systems are powerful tools used in the mechanical design and geometric modeling of products and components.

There are several good reasons for using a CAD system to support the engineering design function:

(1) To increase productivity.
(2) To improve the quality of the design.
(3) To standardize design procedures.
(4) To create a manufacturing database.

【拓展视频】

(5) To eliminate inaccuracies caused by hand-copying drawings and inconsistencies between drawings.

Models in CAD can be classified as being two-dimensional (2D) models, three-dimensional (3D) models, or two-and-a-half-dimensional models $2\frac{1}{2}$D. A 2D model represents a flat part, and a 3D model provides representation of a generalized part shape (Figure 9.1). A $2\frac{1}{2}$D model can be used to represent a part of constant section with no side-wall details. [3]The major advantage of a $2\frac{1}{2}$D model is that it provides some 3D information about a part without needing to create the database of a full 3D model.

After a particular design alternative has been developed, some forms of engineering analysis must often be performed as part of the design process. [4]The analysis may take stress-strain calculations, heat transfer analysis, dynamic simulation, etc. Some examples of software typically offered on CAD systems are mass properties and finite element method (FEM) analysis. Mass properties involve the computation of features of a solid object such as its volume, surface area, weight, and center of gravity. FEM analysis is available on most CAD systems to aid in stress-strain analysis, heat transfer, dynamic characteristics, and other engineering computations. Presently, many CAD systems can automatically generate 2D or 3D FEM meshes essential to FEM analysis.

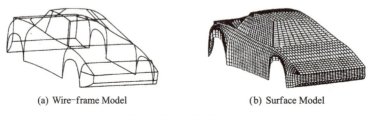

(a) Wire-frame Model　　　　(b) Surface Model

Figure 9.1　3D Models

【拓展视频】

CAM can be defined as computer aided preparation of manufacturing, including decision-making, process and operational planning, software design techniques, artificial intelligence, and manufacturing with different types of automation (NC machine, NC machine centers, and NC flexible manufacturing systems, NC machining cells), and different types of realization (CNC single-unit technology and DNC group technology), as shown in Figure 9.2.

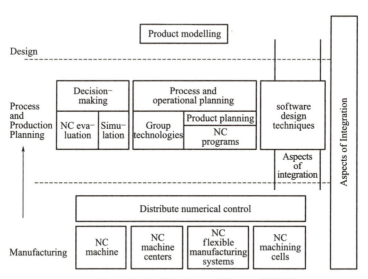

Figure 9.2　The General Scope of CAM

When a design is finalized, manufacturing can begin. Computers play an important role in many aspects of production. Modern shipbuilding fabricates structures from welded steel plates cut from large steel sheets. Computer-controlled flame cutters are often used for this task, and the computer calculates the optimal layout of components to minimize waste metal.

Computer-aided process planning (CAPP) is defined as the function which uses computers to assist the work of process planners. The levels of assistance depend on different strategies employed to implement the system. [5] Lower-level strategies use computers for storage and retrieval of data for process plans constructed manually by process planners, as well as for supplying data used in the new work. In comparison, higher-level strategies use computers to automatically generate process plans for workpieces with simple geometrical shapes. Sometimes a process planner is required to input data or modify plans which do not fit specific production requirements well. The highest-level strategy, the ultimate goal of CAPP, generates process plans by computer, potentially replacing process planners, when the knowledge and expertise of process planning have been incorporated into the computer programs. [6] The database in a CAPP system based on the highest-level strategy will be directly integrated with conjunctive systems, e.g., CAD and CAM. CAPP has been recognized as playing a key role in CIMS (computer integrated manufacturing system).

9.2 Words and Phrases

evolve	v. （使）发展，（使）进展，（使）进化
conceptualization	n. 化为概念，概念化
documentation	n. 文件
inconsistency	n. 不兼容性
numerical control (NC)	数字控制
computer numerical control (CNC)	计算机数字控制
interactive	adj. 交互式的
wire-frame models	线框模型
surface models	表面模型
solid models	实体模型
stress-strain	应力-应变
fabricate	v. 构成，伪造，虚构
incorporate	adj. 合并的，一体化的

9.3 Complex Sentence Analysis

[1] Perhaps the single development that has impacted manufacturing more quickly and significantly than any previous technology is the digital computer.
　　that has impacted manufacturing more quickly and significantly than any previous technology 是定语从句，修饰 development。

[2] Computer aided design (CAD) is defined as the application of computers and graphics software to aid or enhance the product design from conceptualization to

documentation.

① be defined as…：定义成……，定义为……

例如，Teachers are defined as those who do some teaching at school。

② to aid or enhance the product design from conceptualization to documentation 为本句的目的状语；from conceptualization to documentation 为介词短语作定语，修饰 design，其字面解释是：从概念到文件，其实指的就是产品设计过程。

[3] The major advantage of a $2\frac{1}{2}$D model is that it provides some 3D information about a part without needing to create the database of a full 3D model.

① that 引导表语从句。

② without 引导介词短语，在整个句子中充当状语。

例如，I wouldn't have accomplished the designated task without your help.

[4] The analysis may take stress-strain calculations, heat transfer analysis, dynamic simulation, etc. Some examples of software typically offered on CAD systems are mass properties and finite element method (FEM) analysis.

① take：包括，包含。

② finite element method (FEM)：有限元方法，是一种对物体进行物理特性分析的方法，广泛应用在机械学、传热学、电磁学等领域。

[5] Lower-level strategies use computers for storage and retrieval of data for process plans constructed manually by process planners, as well as for supplying data used in new work.

① as well as：也，又；介词性词组；常引导名词、代词或名词性短语。

② for supplying data used in new work 与 for storage and retrieval of data for process plans constructed manually by process planners 作用相当，都是使用计算机的目的。

[6] The database in a CAPP system based on the highest-level strategy will be directly integrated with conjunctive systems, e.g., CAD and CAM.

based on the highest-level strategy 为过去分词短语作定语，修饰 system；based on 在句中作定语。

试比较：

a. Based on the experimental results, it could be inferred that a heat pipe has a strong ability to transfer heat.

b. We do believe the facts based on the experiments.

9.4 Exercise

Translate the following paragraphs

AutoCAD is a computer-aided drafting and design system implemented on a personal computer. It supports a wide range of devices. Device drivers that accompany the system include most digitizers, printer/plotters, video display boards, and plotters available on the market.

AutoCAD supports 2D drafting and 3D wire-frame models and is designed as a single-user CAD package. The drawing elements include lines, poly lines of any width, arcs, circles, faces, and solids. There are many ways to define a drawing element. For example, a circle can be defined by its center and radius, three points, or two end points of its diameter. The system prompts the user for all options, though advanced users can choose to turn off the prompts.

Annotation and dimensioning are also supported. Text and dimension symbols can be placed anywhere on the drawing, at any angle and at any size. A variety of fonts and styles are available as well.

Lesson 10　Engineering Tolerance

10.1　Text

A solid is defined by its surface boundaries. Designers typically specify a component's nominal dimensions to fulfil its requirements. In reality, components cannot be made repeatedly to nominal dimensions due to the surface irregularity and intrinsic surface roughness. [1] Some variability in dimensions must be allowed to ensure manufacture is possible. However, the variability must not be so great that the performance of the assembled parts is impaired. The allowed variability on individual component dimensions is called the tolerance.

【拓展视频】　　【拓展视频】　　【拓展视频】　　【拓展视频】

1. Component Tolerances

Control of dimensions is necessary to ensure assembly and interchangeability of components. Tolerances are specified on critical dimensions affecting clearances and interference fits. One method of specifying tolerances is to state the nominal dimension followed by the permissible variation, such as 40.000 ± 0.003mm. [2] This means the dimension should be machined between 39.997mm and 40.003mm. When the variation occurs on either side of the nominal dimension, the tolerance is called a bilateral tolerance. For a unilateral tolerance, one tolerance is zero, e.g., $40^{+0.006}_{0}$.

Most organizations have general tolerances that apply to dimensions not explicitly specified on a drawing. For machined dimensions, a general tolerance may be ± 0.5mm, allowing a dimension specified as 15.0mm to range between 14.5mm and 15.5mm. Other general tolerances can be applied to angles, drilled and punched holes, castings, forgings, weld beads, and fillets.

[3] When specifying a tolerance for a component, reference can be made to previous drawings or general engineering practice. Tolerances are typically specified in bands as defined in British or ISO standards. Table 10-1 shows examples of tolerance bands and typical applications.

Engineering Tolerance Lesson 10

Table 10 – 1 Examples of Tolerance Bands and Typical Applications

Classifications	Descriptions	Characteristics	ISO codes	Assemblies	Applications
Clearance	Free running fit	Good for large temperature variations, high running speeds, or heavy journal pressures	H9/d9	Noticeable clearance	Multiple bearing shafts; hydraulic position in; cylinders; removable levers; bearing for rollers
	Close running fit	For running on accurate machines and accurate locations at moderate speeds and journal pressures	H8/f7	Clearance	Machine tool main bearings; crankshaft and connecting rod bearings; shaft sleeves; clutch sleeves; guide blocks
	Sliding fit	When parts are not intended to run freely, but must turn and locate accurately	H7/g6	Push fit without noticeable clearance	Push on gear wheels and clutches; connecting rod bearings; indicator pistons
	Location clearance fit	Provides snug fit for locations of stationary parts, but can be freely assembled	H7/h6	Hand pressure with lubrication	Gears; tailstock sleeves; adjusting rings; loose bushes for piston bolts and pipe lines
Transition	Location transition fit	For accurate locations (compromise between clearance and interference fit)	H7/k6	Easily tapped with hammer	Pulleys; clutches; gears; fly wheels; fixed hand wheels and permanent levers
	Location transition fit	For more accurate locations	H7/n6	Needs pressure	Motor shaft armatures; toothed collars on wheels
Interference	Location interference fit	For parts requiring rigidity and alignment with accuracy of locations	H7/p6	Needs pressure	Split journal bearings
	Medium drive fit	For ordinary steel parts or shrink fits on light sections	H7/s6	Needs pressure or temperature differences	Clutch hubs; bearings; bushes in blocks; wheels; connecting rods; bronze collars on grey cast iron hubs

2. Standard Fits for Holes and Shafts

A standard engineering task is to determine tolerances for a cylindrical component, e. g., a shaft, fitting, or rotating inside a corresponding cylindrical component or hole. The tightness of fit depends on the application. [4]For example, a gear located onto a shaft would require a "tight" interference fit, where the diameter of the shaft is slightly greater than the inside diameter of the gear hub to transmit the desired torque. Alternatively, the diameter of a journal bearing must be greater than the diameter of the shaft to allow rotation. [5]Given that it is not economically feasible to manufacture components to exact dimensions, some variability in the sizes of both shaft and hole must be specified. However, the range of variability should not be so large that the operation of the assembly is impaired. Rather than having an infinite variety of tolerance dimensions, national and international standards define bands of tolerances, examples of which are listed in Table 10-1, e. g., H11/c11. Corresponding tables exist to define the tolerance levels for the size of the dimension under consideration. Figure 10.1 shows definitions used in tolerance.

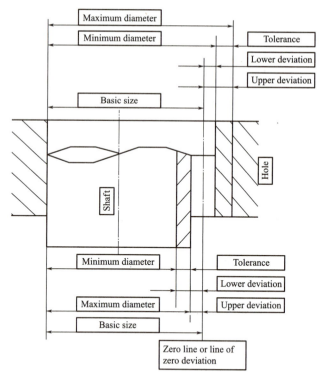

Figure 10.1 Definitions Used in Tolerance

Size: a number expressing the numerical value of a dimension in a particular unit.
Actual size: the size of a part as obtained by measurement.
Limits of size: the maximum and minimum sizes permitted for a feature.
Maximum limit of size: the greater of the two limits of size.

Minimum limit of size: the smaller of the two limits of size.

Basic size: the size by reference to which the limits of size are fixed.

Deviation: the algebraic difference between a size and the corresponding basic size.

Actual deviation: the algebraic difference between the actual size and the corresponding basic size.

Upper deviation: the algebraic difference between the maximum limit of size and the corresponding basic size.

Lower deviation: the algebraic difference between the minimum limit of size and the corresponding basic size.

Tolerance: the difference between the maximum limit of size and the minimum limit of size.

Shaft: the term used by convention to designate all external features of a part.

Hole: the term used by convention to designate all internal features of a part.

10.2 Words and Phrases

tolerance	n. 公差；v. 给机器部件等规定公差
nominal	adj. 公称的，标称的，额定的
intrinsic	adj. 固有的，内在的，本质的
normal distribution	正态分布
weld bead	焊缝
fillet	n. 圆角，倒角
spigot	n. 插销，塞子，阀门
interference fit	干涉配合，过盈配合
broach	n. 拉刀；v. 拉削
gauge	n. （电线等的）直径，（金属板的）厚度，量具
deviation	n. 偏差，偏移

10.3 Complex Sentence Analysis

[1] Some variability in dimensions must be allowed to ensure manufacture is possible.

在该句中，to ensure manufacture is possible 是 Some variability 的主语补足语；manufacture is possible 为 ensure 的宾语从句。

例如，A lot of students will be allowed to do whatever is good for them.

人们允许大多数学生做对他们有益的事情。

[2] This means the dimension should be machined between 39.997mm and 40.003mm.

在该句中，mean 意为"意味着"。实际上，当 mean 作动词时需注意以下两个结构。

① mean to do something：计划做某事、打算做某事，主语通常是人。

例如，I meant to have seen you, but I was too busy at that time.

我原打算去看你的，但是那时实在太忙。

② mean doing something：意味着做某事，主语通常是物。

例如，What he said at the meeting meant his finishing the task given by himself.

他在会上的发言意味着他将独自完成所分配的任务。

[3] When specifying a tolerance for a component, references can be made to previous drawings or general engineering practice.

在该句中，When specifying a tolerance for a component 相当于 When we are specifying a tolerance for a component，由于 specify 的逻辑主语就是主句的主语，且该时间状语的主要谓语动词是 be，因此可以按照英语语法规则将 we are 省去。

[4] For example, a gear located onto a shaft would require a "tight" interference fit, where the diameter of the shaft is slightly greater than the inside diameter of the gear hub to transmit the desired torque.

① 在该句中，located onto a shaft 为过去分词短语作定语，相当于定语从句。

② where the diameter of the shaft is slightly greater than the inside diameter of the gear hub to transmit the desired torque 为非限定性定语从句，修饰 a "tight" interference fit。

例如，The man followed by a lot of students is a professor studying simulation technology.

身后有许多学生跟着的男子是一位从事仿真技术研究的教授。

I like living in Hefei, where a lot of universities are located.

我喜欢生活在合肥，那里有许多大学。

[5] Given that it is not economically feasible to manufacture components to exact dimensions, some variability in the sizes of both the shaft and hole must be specified.

在该句中，Given 同 provided，意为"假定"。

例如，Given that all we want has been prepared, what on earth should we do next?

假定我们所需的一切都准备好了，那么下一步我们究竟该做些什么呢？

10.4　Exercise

Translate the following paragraphs

Computers are used extensively in most engineering functions. Engineering is a profession in which knowledge of the natural sciences is applied with judgment to develop ways of using materials and forces of nature. Typical engineering functions using CAPACS (computer-aided production and control systems) include design, process planning, analysis and optimization, synthesis, evaluation and documentation, simulation, modeling, and quality control planning. Using CAPACS in engineering increases the productivity of engineers and improves the quality of designs.

For example, the application of computers to the engineering design process is performed by a CAD system. Engineers can design and thoroughly test concepts quickly and simply from one workstation. Computers allow engineers to take a concept from its original design through testing to NC output, or a combination of steps in between. They perform complex scientific and engineering computations rapidly with high accuracy, calculate physical properties before actual parts are made, and provide a fast and easy method to create model of even the most complex parts.

The computer has influenced how products are designed, documented, and released for production. As technology develops, engineering operation are becoming more automated, relieving the engineer of many tedious manual calculations.

Lesson 11　A Discussion on Modern Design Optimization

11.1　Text

The integration of optimization techniques with finite element analysis (FEA) and CAD has pronounced effects on the product design process. [1] This integration has the power to reduce design costs by shifting the burden from the engineer to the computer. Furthermore, the mathematical rigor of a properly implemented optimization tool can add confidence to the design process. Generally, an optimization method controls a series of applications, including CAD software as well as FEA automatic solid meshers and analysis processors. This combination allows for shape optimizations on CAD parts or assemblies under a wide range of physical scenarios, including mechanical and thermal effects.

Ideally, there is seamless data exchange via direct memory transfer between CAD and FEA applications without the need for file translation. Furthermore, if associativity between the CAD and FEA software exists, any changes made in the CAD geometry are immediately reflected in the FEA model. In the approach taken by ALGOR, the design optimization process begins before the FEA model is generated. The user simply selects which dimension in the CAD model needs to be optimized and the design criterion, which may include maximum stresses, temperatures, or frequencies. The analysis process appropriate for the design criteria is then performed. The results of the analysis are compared with the design criterion, and, if necessary, without any human intervention, the CAD geometry is updated. [2] Care is taken so that the FEA model is also updated using the principle of associativity, which implies that constraints and loads are preserved from the prior analysis. The new FEA model, including a new high-quality solid mesh, is now analyzed, and the results are again compared with the design criterion. This process is repeated until the design criterion is satisfied. Figure 7.1 shows the procedure of shape optimization.

1. Design Opimization Process

The typical design process involves iterations during which the geometry of the part is altered. In general, each iteration also involves some forms of analysis to obtain viable engineering results. Optimal designs may require a large number of such iterations, each of which is costly, especially if one considers the value of an engineer's time. The principle behind design optimization applications is to relieve the engineer of the laborious tasks by automatically conducting these iterations. At first glance, it may appear that design optimization is a means to replace the engineer and expertise from the design loop. This is certainly not the case because any design optimization application cannot infer what should be optimized, what the design variables are, and the quantities or parameters that can be changed to achieve an optimum design. Thus, design optimization applications are simply another tool available to the engineer. The usefulness of this tool is gauged by its ability to efficiently identify the optimum.

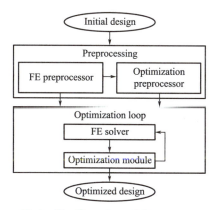

Figure 11.1　The Procedure of Shape Optimization

Design optimization applications tend to be numerically intensive because they must still perform the geometrical and analysis iterations.[3] Fortunately, most design optimization problems can be cast as a mathematical optimization problem for which there exist many efficient solution methods. The drawback to having many methods is that there usually exists an optimum mathematical optimization method for a given problem. This complexity should be remedied by the design optimization application by giving the engineer not only a choice of methods but also a suggestion as to which approach is the most appropriate for their design problem.

A typical optimized quantity is the maximum stress experienced. Typical design variables include geometric quantities, such as the thickness of a particular part. The design of the part or assembly is initiated within a CAD software application. If the component warrants an engineering analysis, the engineer will generally opt to apply FEA to model or simulate its mechanical behavior. The FEA results, such as the maximum stress, can be used to ascertain the validity of the design. During the design process, the engineer may alter parameters or characteristics of the CAD and/or FEA models, including some of the physical dimensions, the material, or how the part or assembly is loaded or constrained. Associativity between the CAD and FEA software should allow the engineer to alter the model in either application and have the other automatically reflect these changes. For example, if the thickness of a part is changed or a hole is added in the CAD software, the FEA model's mesh should automatically reflect those changes. Under most circumstances, engineers will employ linear static FEA to obtain the stresses. This analysis approach has the benefit of yielding a solution for FEA models with many elements in relatively little time. Obviously, linear static FEA has drawbacks as well. For example, significant engineering expertise may be required when estimating the magnitude and direction of loads that are a consequence of motion.

【拓展视频】

【拓展视频】

【拓展视频】

2. Mathematical Theory of Design Optimization

In this section, we focus on the theory underlying some of the mathematical methods employed by design optimization procedures. To describe how the optimization problem arises, we consider a three-step process:

(1) Generation of geometry of part or assembly in CAD.

(2) Creation of FEA model of part or assembly.

(3) Evaluation of results of FEA models.

For now, we limit ourselves to the case of linear static FEA. Therefore, the results are comprised of deflections and stresses at one instance. The manual design process involves all three steps, with the results being used to evaluate whether the design is appropriate. If the design is found inadequate, changes are made to step (1), step (2), or both. It is clear from this description that the output of the FEA results is what should be optimized and that any input to the CAD or FEA models can be viewed as a design variable. A design optimization algorithm conducts many FEA operations, each one with a different set of values for the design parameters. Before the manual design approach can be transformed into a design optimization algorithm, there must be associativity between the CAD and FEA applications. The rational behind this requirement is best explained using an example. Consider the initial design stage when the engineer applies constraints on a particular surface of the FEA model; it can be safely assumed that this surface coincides with a surface in the CAD model. Now, if the design optimization algorithm decides to alter the geometry of the CAD surface, then the FEA model must automatically reflect these changes and apply the constraints on the new representation of this surface. Thus, associativity is required to achieve this automatic communication between the CAD and FEA models. Having defined the design optimization problem for mechanical systems, we now describe the mathematics used to solve these problems.

Most optimization problems are made up of three basic components:

(1) An objective function which we want to minimize (or maximize). For instance, in

designing an automobile panel, we might want to minimize the stress in a particular region.

(2) A set of design variables that affect the value of the objective function. In the automobile panel design problem, the variables used define the geometry and material of the panel.

(3) A set of constraints that allow the design variables to have certain values but exclude others. In the automobile panel design problem, we would probably want to limit its weight.

It is possible to develop an optimization problem without constraints. Some may argue that almost all problems have some forms of constraint. For instance, the thickness of the automotive panel cannot be negative. Although in practice, answers that make good sense in terms of the underlying physics, such as a positive thickness, can often be obtained without enforcing constraints on the design variables.

3. Benefits and Drawbacks

The elimination or reduction of repetitive manual tasks has driven the development of many software applications. Automatic design optimization is one of the latest applications used to reduce man-hours, possibly increasing computational efforts. It is even possible for an automatic design optimization scheme to require less computational efforts than a manual approach, as the mathematical rigor of these schemes may be more efficient than a human-based solution. However, these schemes do not replace human intuition, which can occasionally significantly shorten the design cycle.[4] One definite advantage of automated methods over manual approaches is that software applications, if implemented correctly, should consider all viable possibilities. No variable combination of the design parameters is left unconsidered, ensuring designs obtained using design optimization software are accurate within the method's resolution.

11.2 Words and Phrases

optimization	n. 最佳化，最优化
finite element analysis (FEA)	有限元分析
computer aided design (CAD)	计算机辅助设计
burden	n. 担子，负担，责任，义务
rigor	n. 严格，严密，精确
mesh	n. 网孔，网格，网状物
scenario	n. 方案，情况
seamless	adj. 无缝的，无伤痕的
criterion	n. 标准，规范，准则，判据

iteration	*n.* 反复，迭代
geometry	*n.* 几何学，几何图形（形状），表面形状
gauge	*v.* 判断，测试，测定，测量
intensive	*adj.* 强化的，加强的
cast	*v.* 计算，派（角色），分类整理
assembly	*n.* 组合，装配，部件，汇编
variable	*n.* 可变物，变量；*adj.* 可变的，变量的
warrant	*v.* 成为……的证据，保证，证明……是正确的
yield	*v.* 产出，产生，提供，给予，得出
expertise	*n.* 专家的意见，专门知识，经验，专家
underlying	*adj.* 基础的，根本的，在下面的
linear	*adj.* 线的，直线的，线性的
function	*n.* 功能，作用，职责，函数
constraint	*n.* 约束，强制，局促
impetus	*n.* 推动力，促进，刺激，激励
scheme	*n.* 计划，阴谋，方案，图解

11.3　Complex Sentence Analysis

[1] This integration has the power to reduce design costs by shifting the burden from the engineer to the computer.
① power：能力；has the power to：能够。
② 与 ability 相比，power 更强调本能、智能和体能。
试比较：
  a. Some animals have the power to see in the dark.
  b. He has a strong ability to deal with the business.

[2] Care is taken so that the FEA model is also updated using the principle of associativity, which implies that constraints and loads are preserved from the prior analysis.
① so that 引导状语从句。
② which 引导非限定性定语从句，which 代表前面这个句子整体。
③ 此外，上述句子可写成以下两种形式：
Care is taken so that the FEA model is also updated using the principle of associativity, as implies that constraints and loads are preserved from the prior analysis.
As implies that constraints and loads are preserved from the prior analysis, care is taken so that the FEA model is also updated using the principle of associativity.

[3] Fortunately, most design optimization problems can be cast as a mathematical

optimization problem for which there exist many efficient solution methods.
① cast：派（角色）；be cast as：看成。
② for which…为介词＋which 引导定语从句，修饰 problem。
试比较：

At an instant, the energy in the control volume includes the rate at which thermal and mechanical energy enters and leaves through the control surface.

[4] One definite advantage of automated methods over manual approaches is that software applications, if implemented correctly, should consider all variable possibilities.
① advantage…over…：与……相比的优越性。
② that 引导表语从句，其中 if implemented correctly 是插入语。

11.4　Exercise

Translate the following paragraphs

Optimization is concerned with finding the best possible solution, formally referred to as the optimal solution, to a particular problem (e.g., a design problem). The term "optimization" is used loosely in general speech, but here it has a precise meaning: the action of finding the best possible solution to a problem as defined by an unambiguous criterion.

Generally, there is more than one solution to a design problem, and the first solution is not necessarily the best. The need for optimization is inherent in the design process. A mathematical theory of optimization has become highly developed and is applied to design where design functions can be expressed by mathematical equations or finite element computer modeling. Optimization techniques may require considerable knowledge and mathematical skills to select the appropriate optimization techniques and work through to a solution. The growing acceptance of the Taguchi method comes from its applicability to a wide variety of problems with a methodology that is not highly mathematical.

By optimal design, we mean the best of all feasible designs. Optimization is the process of maximizing a desired quantity or minimizing an undesired one. Optimization theory is the body of mathematics that deals with the properties of maxima and minima and how to find them numerically.

Lesson 12 Using Dynamic Simulation in the Development of Construction Machinery

12.1 Text

【拓展视频】

[1] The general motives for "virtual prototyping" are probably familiar to all engineers: stricter legal requirements (e.g., with regard to exhaust emissions and sound) and tougher customer demands (e.g., with regard to performance and handling) lead to more advanced, complex systems which are harder to optimize. With traditional methods, development costs more and takes longer. In contrast, increased competition demands lower development costs and shorter project times.

Virtual prototyping has been generally adopted in the vehicle industry as a major step toward solving this conflict, both on the consumer side (cars) and on the commercial side (trucks and buses, as shown in Figure 12.1). Starting with simulation of sub-systems, the state-of-the-art now includes simulation of complete vehicles, mostly for evaluation of handling, comfort, and durability, but also for crash tests.

Figure 12.1 Multi-body Model of a Volvo L220E Wheel Loader

[2] One reason for the off-road equipment industry lagging behind is the size of these companies. Being significantly smaller, broad investments in the latest CAE tools (together with the necessary training) take longer to amortize. The other, and probably more important reason, is that the products are very different to those of the on-road vehicle industry—not only geometrically (size) but also topologically (sub-systems of various domains and their interconnections).

Recently, cases have been published where complete machines were simulated for evaluating

Using Dynamic Simulation in the Development of Construction Machinery Lesson 12

the simulation technique itself, subsystems, comfort-related aspects, or durability. This section will also deal with the dynamic simulation of complete machines, but for analysis and optimization of overall performance and related aspects. The focus will be on wheel loaders with hydrodynamic transmissions, but most findings (and questions) will also be applicable to the off-road machinery.

【拓展视频】

1. Design Process and Visualization

The aim of the present project is the evolution of the current product development process, rather than a revolution by means of "Design Science". The research question is how to augment the existing design process with dynamic simulation. [3] As mentioned before, the focus is on analysis and optimization of overall performance and related aspects.

The revised design process must fulfil the following non-optional targets:

(1) Products of at least equal high performance, efficiency, and operability.

(2) With increased robustness.

(3) In a shorter time.

(4) At a lower development cost.

In an earlier project, a valuable lesson was learned: speed matters when it comes to iterations, especially in the concept phase. When Volvo's old loading unit calculation program was replaced with a more modern version, it was done with a proprietary simulation system based on a multi-body simulation system and a modern database. The development was done in-house. This new simulation system proved to be more flexible, more accurate, and especially more efficient for the user, except for some pre-study engineers who used the superior speed of the old calculation program to brute force optimize loading unit geometries. Since the new system obtains results by multi-body simulation, rather than calculation of hard-coded explicit equations, one run takes a couple of seconds longer than with the old program. Brute-force optimizations of the old type are no longer time-efficient. If this had been known before, i. e. , if it had been included in the project targets, a special downscaled version, less accurate but faster, could have been developed. The introduction of the new system forced those pre-study engineers to abandon a time-efficient technique that worked well.

A similar risk can be seen with this research project: the current initial static calculation loop (Figure 12. 2) is fast and reasonably accurate. The shortcoming today is that the dynamic behaviour of the complete machine is first evaluated by testing a functional prototype, followed by testing a "real" prototype. Therefore, a moderately revised process is proposed.

As practiced today, the process starts with feeding the product targets into the initial static calculation loop. If no satisfactory solution can be found, the targets need to be revised. The next step is described as "dynamically augmented, static calculations". Here, as

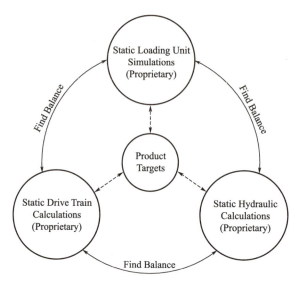

Figure 12.2　The Current Initial static Calculation Loop

well as in the next step "dynamic simulation of complete machine", non-fulfilled product targets do not necessarily lead to a reiteration, as long as the deviation can be approved. Since product targets will probably never cover all product properties (including dynamic behaviour), this checkpoint will give the whole process with some flexibility. [4] Only if the deviation is too high will a new static calculation loop need to be started, with the original product targets as input. If no alternative layout candidate can be calculated, the product targets need to be revised. Since simulating complete machines can take a long time, a quick check on whether the solution from the initial loop is dynamically feasible would be of great help. This is supposed to be done in the second step, "dynamically augmented, static calculations". Concrete methods still need to be developed. However, one example can be given for machines with hydrodynamic transmissions and LS-hydraulics: a critical phase in a so-called "short loader cycle" (or V-cycle) is when the machine, coming backwards with a full bucket, changes direction towards the load receiver (e.g., an articulated hauler or a dump truck). During that time, there is a close interaction between the main subsystems:

(1) To reverse the machine, the operator lowers the engine speed to prevent jerky gear shifting and premature transmission wear. Less torque is available at lower speeds, and the engine response is worse, mainly due to the turbocharger's inertia and smoke limiter settings.

(2) When switching gears from reverse to forward, the loader is still rolling backwards. This forces an abrupt change in the rotational direction of the torque converter's turbine wheel, greatly increasing the slip and leading to a sudden increase in torque demand from the engine.

Using Dynamic Simulation in the Development of Construction Machinery　Lesson 12

(3) Some operators do not stop lifting the loading unit while reversing, requiring high oil flow throughout the process. The oil flow is proportional to the hydraulic pump's displacement and shaft speed, and the load-sensing pumps are directly connected to the engine crankshaft. Thus, with lower engine speeds and high demand of oil flow, the displacement reaches maximum. The torque demanded by the pump is proportional to its displacement and hydraulic pressure. Since the loader's bucket is full and due to the loading unit's geometry, hydraulic pressure is high. With displacement at maximum, this also leads to increased torque demand.

Both the drivetrain and hydraulics suddenly apply a higher load to an already weakened engine. All depends on the time scale of these three concurrent phenomena, which is why a satisfactory answer can only be given by a detailed dynamic simulation or testing a real machine, i. e. , a functional prototype. However, an approximate and less time-consuming first approach is possible. Given the loader speed when switching gears from reverse to forward and given the engine speed at that time, the maximum slip between the pump wheel and the turbine wheel of the torque converter can be calculated, and thus the maximum demanded engine torque (using the torque converter's specifications). In the worst-case scenario, hydraulic pressure and pump displacement can be assumed to be maximum. Together with the engine speed, this gives the second engine torque demand. If the sum of both torque demands is larger than the available steady-state engine torque at that speed, the proposed system layout will almost certainly lead to dynamic problems. If the available steady-state engine torque is considerably larger than the sum of both torque demands, the system will most probably function as intended. [5] To check the case in between, it is important to consider that due to factors such as turbocharger's inertia and smoke limiter mentioned before, an accelerating engine seldom has full steady-state torque at lower engine speeds. Therefore, checking against the static torque curve might give a false sense of security.

What is needed is a dynamic engine torque curve, measured at a typical less gap between static and dynamic torque at lower acceleration (the bold dashed curve in Figure 12.3).

Assuming the engine is released to low idle during reversing (worst-case), the acceleration rate is simply obtainable as the speed difference from the engine's max torque speed to low idle, divided by a typical time for the reversal phase. Using a text bench with an electro magnetic brake, an engine run-up with a forced acceleration as described above can be performed. The available torque during that phase equals the necessary braking torque.

In Figure 12.3, the dashed curves represent a test where the engine was allowed to accelerate less. This theoretically generates more available torque because the rotating parts consume less torque for acceleration, and there is more time for turbo boost pressure to be built up. While the engine torque

【拓展视频】

does not increase faster with less forced acceleration, more torque is available relative to engine speed. This is expressed as a smaller gap between the static and dynamic torque curves for the slower acceleration. If the engine were allowed to accelerate sufficiently slowly enough, the dynamic torque curve (bold line) would follow the static one (thin line).

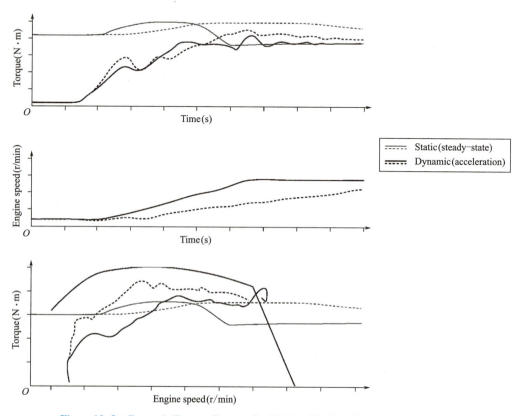

Figure 12.3　Dynamic Torque Curves of a Modern Turbo—Charged Diesel Engine

To quickly identify system layouts that might lead to dynamic problems, more checking methods and refinements of existing rules of thumb need to be developed.

Taking a critical look at the proposed revised design process, even if it has been labeled as "evolutionary rather than revolutionary", it implies a bigger change than might be obvious at the first glance. With its introduction, functional prototypes are only built to verify system layouts that have passed both steps that check the dynamic behavior. These steps are given the great responsibility of deciding whether a certain system layout can be further pursued or should be sent back for reiteration. The worst scenario would be rejecting cases that would actually work in reality, but nobody will ever know since only approved cases enter the functional prototype phase. To avoid this and build confidence, there has to be a smooth transition between current practice and the proposed process.

2. Conclusion and Outlook

This section presented the joint research by Volvo Wheel Loaders and Linköping University on simulation of complete machines for analysis and optimization of overall performance. The motivation on the side of the industrial partner is to develop products of equally high performance, efficiency, and operability, but with more robustness regarding these aspects, in a shorter time and at a lower total development cost. A revised product development process (with regard to the research topic) has been proposed. Examples of areas for future research have also been presented. Research in the immediate future will focus on a definition of operability (including quantification), as well as practical simulation problems (such as defining modules and physically motivated and intuitive interfaces between them).

In a longer perspective, a measure of complexity needs to be developed. Research will be done on how to optimize operability, which will also include a look at control strategies.

12.2 Words and Phrases

dynamic	*adj.* 动力的，动力学的，动态的
dynamic simulation	动态仿真
prototype	*n.* 原型，模型，典型，榜样，样机
virtual prototyping	虚拟原型机
off-road	*adj.* 道路之外的，越野的
lag	*v.* 缓缓而行，滞后；*v.* 落后于，押往监狱，加上外套
amortization	*n.* 摊销，分期偿付，分期偿还
topologically	*adv.* 拓扑地
hydrodynamic	*adj.* 水力的，水压的，液力的，流体动力学的
augment	*v.* 增加，增大；*n.* 增加
operability	*n.* 操作性能，可操作性
robustness	*n.* 强壮，雄壮，健全，耐用，坚固
proprietary	*adj.* 专利的，独占的，有财产的，私有的
downscale	*v.* 缩减……规模，按比例缩减
reiteration	*n.* 重复，反复
checkpoint	*n.* 关卡，检查点
jerky	*adj.* 急拉的，急动的
concurrent	*adj.* 同时，兼，并行地
scenario	*n.* 游戏的关，某特定情节
intuitive	*adj.* 直觉的，本能的，天生的

perspective　　　　　　　　　　　　　n. 透视图，远景，前途，观点，看法，观察

12.3 Complex Sentence Analysis

[1] The general motives for "virtual prototyping" are probably familiar to all engineers.

be familiar to = be known to：（某人或某物）对……是熟悉的。

例如，The food and the climate here are familiar to me.

我对这里的饮食和气候比较熟悉。

试比较：be familiar with，（某人）熟悉……

She is still not familiar with the work.

[2] One reason for the off-road equipment industry lagging behind is the size of these companies.

在该句中，lagging behind 作定语，修饰 industry，意为"落后、滞后"。

[3] As mentioned before, the focus is on analysis and optimization of overall performance and related aspects.

在该句中，as 替代 the focus is on analysis and optimization of overall performance and related aspects。

[4] Only if the deviation is too high will a new static calculation loop need to be started, with the original product targets as input.

当 only 引导副词、副词性词组和状语句子时，后面跟的主句一般需倒装。

试比较：

a. Only then did I realize made such a big mistake.

只有在那时，我才认识到我犯了一个多么大的错误。

b. Only when one loses health does he know its value.

只有当人们身体不好时才会认识到健康的重要性。

[5] To check the case in between, it is important to consider that due to factors such as turbocharger's inertia and smoke limiter mentioned before, an accelerating engine seldom has full steady-state torque at lower engine speeds.

in between：在中间，在……之间；挡路。

12.4 Exercise

Translate the following paragraph

As shown earlier, off-road equipment consists of systems from various domains, and most of them need to be taken into account when simulating performance of complete machines, which is usually a collaborative activity. As noted by many researchers, engineers have often already chosen a domain-specific simulation program that they are familiar with. Instead of forcing a migration to one monolithic simulation system that can be used in several domains (but offers only limited functionality in the individual domain), a better approach is to couple the specialized and single-domain tools. This has the advantage that both pre-processing and post-processing are done decentralized in the engineer's domain-specific tools. J. Larsson develops a technique for such co-simulation and applies it to a model of a complete wheel loader. The present research project will use this approach to further develop it.

Lesson 13　Introduction to Heat Pipe Technology in Machining Process

13.1　Text

【拓展视频】

【拓展视频】

In any machining process, most of the input energy is converted into heat in the cutting zone, resulting in increased temperatures of tool and workpiece. Elevated temperatures can significantly shorten tool life. Excessive heat in the tool and workpiece can lead to thermal distortion and poor dimensional control. In addition, high tool temperatures promote the formation of BUE (built-up edge) on the tool tip.

[1] In a drilling process, tool temperatures are particularly important because the chips, which absorb much of the cutting energy, are generated in a confined space and remain in contact with the tool for a relatively long time compared to other machining operations. As a result, drill temperatures are much higher than in other processes under similar conditions. The most common cooling method is the use of cutting fluids flooding through the cutting zone. In machining operations, three types of cutting fluids are commonly used: oil with additives such as sulfur, chlorine, phosphorus, emulsions, and synthetics containing inorganic and other chemicals. Used coolant from machining processes is harmful to both environment and human health. Chemical substances that provide lubrication in the machining process are toxic if released into soil and water. These chemical substances in coolant can cause serious health problems for workers exposed to the coolant in both liquid and mist forms. Researchers have proposed replacing conventional fluids with cryogenic fluids such as liquid nitrogen or carbon dioxide. Although this method shows promise in increasing tool life and eliminates the need for cutting fluid removal and disposal, many technical issues, including safety, remain unresolved. Recently, there has been a strong global trend towards minimizing cutting fluids because they are primary sources of industrial pollution. [2] To enable a machining process to run dry, an effective cooling method, other than flooding with a coolant, is desirable to remove the heat accumulated in the drill tool.

Heat pipes have been proved to be an effective alternative to conventional methods for removing heat from a drill tip, allowing drilling operations to be carried out in a dry and "green" manner. The components of a heat pipe are a sealed container (pipe wall and end caps), a wick structure, and a small amount of working fluid in equilibrium with its own vapor. Figure 13.1 illustrates physical configuration of the heat pipe. Typically, a heat pipe can be divided into three sections: evaporator section, adiabatic (transport) section, and condenser section. The external heat load on the evaporator section causes the working fluid to vaporize.

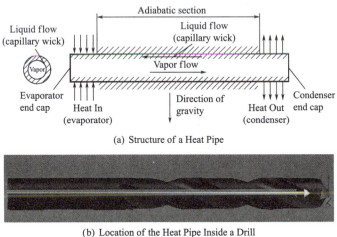

(a) Structure of a Heat Pipe

(b) Location of the Heat Pipe Inside a Drill

Figure 13.1 Physical Configuration of the Heat Pipe

The resulting vapor pressure drives the vapor through the adiabatic section to the condenser section, where the vapor condenses, releasing its latent heat of vaporization to the cooler environment. The condensed working fluid is then pumped back by capillary pressure generated by the meniscus in the wick structure. Heat transport can be continuous as long as sufficient capillary pressure is generated to drive the condensed liquid back to the evaporator.

[3] From a heat transfer point of view, researchers demonstrated the feasibility and effectiveness of heat pipes in drilling operations. Key findings include:

(1) Both numerical studies and initial experiments show that using a heat pipe inside the drill significantly reduces the temperature field (Figure 13.2).

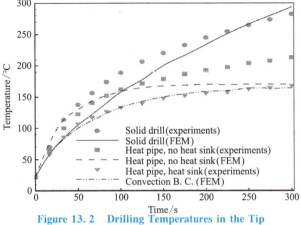

Figure 13.2 Drilling Temperatures in the Tip

(2) The closer the heat pipe is to the tip, the more effective it removes the heat. The stresses in the drill do not increase significantly with longer heat pipes, but manufacturing constraints limit the pipe's length (Figure 13.3).

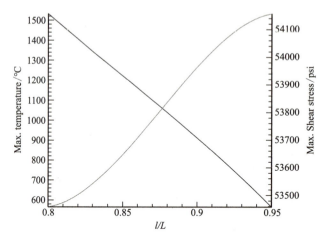

Figure 13.3　Effect of Ratio heat pipe's length (l) to drill's length (L)

(3) The diameter of the heat pipe affects stress levels more than the maximum temperatures reached. While the diameter does not impact the maximum temperature, it influences the temperature distribution (Figure 13.4).

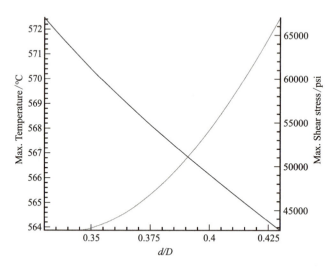

Figure 13.4　Effect of Ratio Heat Pipe's Diameter (d) to Dill's Diameter (D)

13.2　Words and Phrases

cutting zone　　　　　　　　切削区域
workpiece　　　　　　　　　$n.$ 工件

Introduction to Heat Pipe Technology in Machining Process Lesson 13

thermal distortion	热变形，热扭曲
cutting fluid	切削液
additive	*adj.* 附加的，加成的，添加的；*n.* 添加剂
sulfur	*n.* [化] 硫黄；*vt.* 用硫黄处理
chlorine	*n.* [化] 氯
phosphorus	*n.* [化] 磷
emulsion	*n.* 乳状液，[医] 乳剂，[摄] 感光乳剂
cryogenic	*adj.* 低温学的，低温的，低温存储的
meniscus	*n.* 新月，半月板，（液柱的）弯月面

13.3 Complex Sentence Analysis

[1] In a drilling process, tool temperatures are particularly important because the chips, which absorb much of the cutting energy, are generated in a confined space and remain in contact with the tool for a relatively long time compared with any other machining operation.

① which absorb much of the cutting energy 引导非限制性定语从句，修饰 chips。

② compared with any other machining operations 为过去分词作定语，修饰 time。

试比较：

The approaches proved by the experiments in the lab have been widely employed in industries.

用实验室实验证明的方法在工业中得到了广泛的应用。

[2] To enable a machining process to run dry, an effective cooling method, other than flooding with a coolant, is desirable to remove the heat accumulated in the drill tool.

① machining process to run dry：干式机械加工过程，run 为连系动词，dry 作表语。

② other than：与……不同。

试比较：

You can't get there other than by swimming.

[3] From a heat transfer point of view, researchers demonstrated the feasibility and effectiveness of heat pipes in drilling operations.

From a heat transfer point of view：从传热的角度。From…point of view 表示"从……的角度"。

试比较：

From a static point of view：从静态的角度；From dynamic point of view：从动态的角度。

13.4 Exercise

Translate the following paragraph

The operation of a heat pipe can be analyzed using a fundamental thermodynamic cycle for the working fluid, in which thermal energy is converted into kinetic energy. This cycle can be studied using a temperature-entropy diagram. According to the second law of thenmady namics, converting thermal energy into kinetic energy involves heat rejection at a lower temperature, resulting in an efficiency of less than 100%. A heat pipe cannot be completely isothermal because this would violate the second law. If the thermal conductivity is measured based on the temperature difference in the system with the same amount of heat transferred, the thermal conductivity of a heat pipe system is much larger than that of pure conduction in a solid rod with the same dimensions. This is because of the heat transfer mechanism involving the latent heat of evaporation and condensation. Although a thermodynamic approach provides insight into the performance, this analysis is very limited, and we need to appeal to heat transfer and fluid mechanics to solve the problem quantitatively. Fluid flow analysis is used to describe the axial liquid pressure drop in the wick structure, the maximum capillary pumping head, and vapor flow in the vapor channel. Combined with heat transfer analysis, this approach models heat transfer and fluid flow in the evaporator and condenser, as well as forced convection in the vapor channel. These detailed analysis help identify optimum operation conditions, improving heat pipe cooling performance through design enhancements.

Lesson 14　Introduction to Material Forming

14.1　Text

1. Material Forming Processes as a System

The term "material forming" refers to a group of manufacturing methods by which the given shape of a workpiece (a solid body) is converted into another shape without changing the mass or the material composition of the workpiece.

Material forming is used synonymously with deformation or deforming and comprises the methods in group II of the manufacturing process classification shown below. The manufacturing processes are divided into six main groups:

Group I (primary forming): Original creation of a shape from the molten or gaseous state or from solid particles of undefined shape, that is, creating cohesion between particles of the material.

Group II (deforming): Converting a given shape of a solid body into another shape without changing the mass or the material composition, that is, maintaining cohesion.

Group III (separating): Machining or removal of material, that is, destroying cohesion.

Group IV (joining): Uniting of individual workpieces to form subassemblies, filling and impregnating workpieces, that is, increasing cohesion between several workpieces.

Group V (coating): Application of thin layers to a workpiece, for example, galvanizing, painting, coating with plastic foils, that is, creating cohesion between substrate and coating.

Group VI: (changing the material properties) Deliberately changing the properties of the workpiece to achieve optimum characteristics at a particular point in the manufacturing process.[1] These methods include changing the orientation of microparticles as well as their introduction and removal, such as by diffusion, that is, rearranging, adding, or removing particles.

In manufacturing technology, particularly in groups I to IV, we continually face with the challenge of how to manufacture a particular technical product the most economically, with specific tolerance requirements, surface structure, and material properties.

2. Cold-Forming and Hot-Forming

[2]By applying a stress that exceeds the original yield strength of metallic material,

we strain-harden or cold-work the material while simultaneously deforming it. This forms the basis for many manufacturing techniques, such as wire drawing. Figure 14.1 illustrates several manufacturing processes that use both cold-working and hot-working processes.

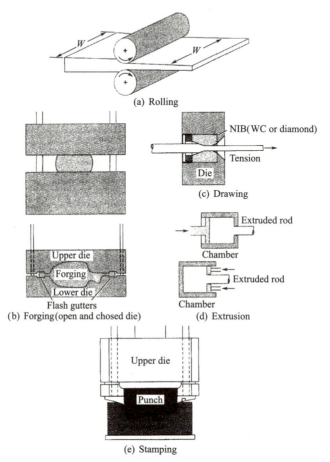

Figure 14.1 Several Manufacturing Processes that Use Both Cold-working and Hot-working Processes

Many techniques simutaneously shape and strengthen material by cold-working. For example, rolling [Figure 14.1(a)] produces metal plates and sheets. Forging [Figure 14.1 (b)] deforms material into a die cavity, producing complex shapes like automotive crankshafts or connecting rods. In drawing [Figure 14.1(c)], a metallic rod is pulled through a die to produce wire or fiber. In extrusion [Figure 14.1(d)], material is pushed through a die to form products with uniform cross-sections, including rods, tubes, or aluminum trims for doors or windows. Deep drawing forms the body of aluminum beverage cans. Stamping [Figure 14.1(e)], bending, and other processes shape materials. Cold-working effectively shapes metallic materials while increasing their strengths. It provides excellent dimensional tolerances and surface finishes. Note that many processes, such as rolling, can be conducted using both cold-working and hot-working.

We can also deform metal into useful shapes by hot-working rather than cold-working. As described previously, hot-working involves plastically deforming metallic material above the recrystallization temperature. Hot-working is well-suited for forming large parts, as metal has low yield strength and high ductility at elevated temperatures. In addition, HCP metals like magnesium have more slip systems at hot-working temperature; the high ductility allows larger deformation than cold-working permits. For example, a very thick plate can be reduced to a thin sheet in a continuous series of operations.

An advantage of hot-working is that materials imperfection can be eliminated during the process. Some imperfection in the original materials may be reduced or minimized; gas pores can be closed and welded shut; composition differences in the metal can be reduced; and the structure of metals can be refined and controlled by recrystallization. Therefore, mechanical and physical properties of metals can be significantly improved.

3. Principles of Plastic Forming

Plasticity theory is the foundation for the numerical treatment of metal forming processes. Materials science and metallurgy explain the origins of the plastic state of metallic bodies and its dependence on various parameters, such as process speed, prior history, temperature, and so on. The older plasticity theory deals with the calculation of stresses, forces, and deformation.

Plasticity theory is based on macroscopically observed phenomena, or the properties of materials observed and measured directly in deformation processes, such as tension and compression tests. This leads to the following simple description of the plastic state.

[3] Plasticity is the capacity of a material to change shapes permanently under the action of forces when the stress reaches a material-dependent critical magnitude called yield strength or initial flow stress. As seen in tension test, when stress is below the yield strength, deformation disappears upon unloading, and the material behaves elastically. If stress exceeds the yield strength, permanent deformation forms. Upon unloading, the workpiece has a form different from its initial one. It is said to have been plastically or permanently deformed, or if a definite final shape was sought, it has been transformed. Materials which behave in an elastic-plastic manner can, after permanently deformed, be loaded again until flow stress is reached without additional permanent deformation. This increase in flow stress as a result of prior deformation is called strain hardening or work hardening.

Work hardening can be balanced by the dynamic softening processes of recovery and recrystallization. The original cold-worked microstructure consists of deformed grains with many tangled dislocations. When we first heat the metal, the additional thermal energy allows dislocations to move and form the boundaries of a polygonized subgrain structure.

The dislocation density, however, remains virtually unchanged. This low-temperature treatment, which removes residual stresses from cold-working without changing dislocation density, is called recovery.

When a cold-worked metallic material is heated above a certain temperature (recrystallization temperature), rapid recovery eliminates residual stresses and produces the polygonized dislocation structure. New small grains nucleate at the cell boundaries of the polygonized structure, eliminating most dislocations. With fewer dislocations, the recrystallized metal has low strength and high ductility. The process of forming new grains by heat treating a cold-worked material is known as recrystallization. Recrystallization may be followed by grain growth if the temperature is sufficiently high.

4. Methods Used in Material Forming

The following classification of deformation methods into five groups is based mainly on the important differences in effective stresses. No simple descriptions of stress states are possible since, depending on the kind of operation, different stress states may occur simultaneously, or they may change during the course of the deforming operation. Therefore, the predominant stresses are chosen with the classification criteria. The five groups of material-forming processes may then be defined as follows:

(1) Compressive forming (forming under compressive stresses): The German standard covers the deformation of a solid body in which the plastic state is achieved mainly by uniaxial or multiaxial compressive loading.

(2) Combined tensile and compressive forming (forming under combined tensile and compressive stresses).

(3) Tensile forming (forming under tensile stresses).

(4) Forming by bending (forming by means of bending stresses).

(5) Forming by shearing (forming under shearing stresses).

14.2　Words and Phrases

synonymously	*adv.* 同义地
cohesion	*n.* 结合，凝聚，[物理] 内聚力结合
subassembly	*n.* 部件，组件
impregnate	*v.* 使充满，使怀孕，注入，灌输；*adj.* 充满的，怀孕的
stamping	*n.* 冲压，压印，模压，模锻
rolling	*n.* 轧制，旋转，翻滚；*adj.* 旋转的，转动的，摇摆的，起伏的
forging	*n.* 锻造，锻炼，锻件

Introduction to Material Forming Lesson 14

crankshaft	n. 曲轴，机轴
cavity	n. 洞，空穴，型腔
connecting rod	连杆
drawing	n. 拉拔，制图，图画
extrusion	n. 挤压，挤出，推出
deep drawing	拉深
stretch forming	拉伸变形
bending	n. 弯曲（变形）
HCP metals	密排六方金属
imperfection	n. 缺陷，不完整性，非理想性
gas pores	气孔
galvanise	v. & n. 电镀
foil	n. 箔，金属薄片，烘托，衬托；v. 贴箔于……，衬托，阻止
substrate	n.（＝substratum）底层，下层，[地]底土层，基础
deliberately	adv. 有目的地，故意地
metallurgy	n. 冶金，冶金学，冶金术
macroscopically	adv. 宏观上，宏观地
recovery	n. 回复
recrystallization	n. 再结晶
tangled	adj. 紊乱的，复杂的
dislocation	n. 位错
polygonized subgrain structure	多边形亚晶结构
grain growth	晶粒长大
predominant	adj. 卓越的，支配的，主要的，突出的，有影响的
shearing stress	剪应力

14.3 Complex Sentence Analysis

［1］These methods include changing the orientation of micro-particles as well as their introduction and removal，such as by diffusion，that is，rearranging，adding，or removing particles.
① as well as：和，还有。
② rearranging：重排。

［2］By applying a stress that exceeds the original yield strength of metallic material，we strain-hardened or cold-work the material while simultaneously deforming it.

① strain-hardened：应变硬化。
② while simultaneously：而同时。
[3] Plasticity is the capacity of a material to change shapes permanently under the action of forces when the stress reaches a material-dependent critical magnitude called yield strength or initial flow stress.
① under the action of…：在……作用下。
② critical magnitude：临界值。
③ flow stress：流变应力。

14.4 Exercise

Translate the following paragraphs

The ceramic group of cutting tools represents the most recent development in cutting tool materials. They consist mainly of sintered oxides, usually aluminum oxide, and are almost invariably in the form of clamped tips. Because of the comparative cheapness of ceramic tips and the difficulty of grinding them without causing thermal cracking, they are made as throw-away inserts.

Ceramic tools are a post-war introduction and are not yet in general factory use. Their most likely applications are in cutting metal at very high speeds, beyond the limits possible with carbide tools. Ceramics resist the formation of a built-up edge and consequently produce good surface finishes. Since the present generation of machine tools is designed with only sufficient power to exploit carbide tooling, it is likely that, for the time being, ceramics will be restricted to high-speed finish machining where there is sufficient power available for the light cuts taken. The extreme brittleness of ceramic tools has largely limited their use to continuous cuts, although their use in milling is now possible.

As they are poorer conductors of heat than carbides, temperatures at the rake face are higher than in carbide tools, although the friction force is usually lower. To strengthen the cutting edge and consequently improve the life of the ceramic tool, a small chamfer or radius is often stoned at the cutting edge, although this increases power consumption.

Lesson 15 Material Forming Processes

15.1 Text

In this section, a short description of the process examples about material forming will be given. However, assembly and joining processes are not described here.

1. Forging

Forging can be characterized as mass conserving, involving the solid state of work material (metal), and a mechanical primary basic process of plastic deformation.[1] Technically, forging is defined as the process of enhancing metal utility by shaping, refining, and improving its mechanical properties through controlled plastic deformation under impact or pressure.

【拓展视频】
【拓展视频】

A wide variety of forging processes are used, and Figure 15.1 (a) shows the most common: drop forging. The metal is heated to a suitable working temperature and placed in the lower die cavity. The upper die is then lowered so the metal is forced to fill the cavity. [2] Excess material is squeezed out between the die faces at the periphery as flash, which is removed in a later trimming process. When the term forging is used, it usually means hot forging. Cold forging has several specialized names. Material loss in forging processes is usually quite small. Normally, forged components require some subsequent machining, as the tolerances and surfaces obtainable are not usually satisfactory for a finished product. Forging machines include drop hammers and forging presses with mechanical or hydraulic drives. These machines involve simple translatory motions.

2. Rolling

Rolling can be characterized as mass conserving, involving the solid state of material, and a mechanical primary basic process of plastic deformation. Rolling is extensively used in manufacturing plates, sheets, structural beams, and more. Figure 15.1 (b) shows the rolling of plates or sheets. An ingot produced in casting is reduced in thickness over sereval stages, usually while hot. Since the width of the work material is kept constant, its length increases according to the reductions. After the last hot-rolling stage, a final stage is carried out cold to improve surface quality and tolerances and increase strength. In rolling, the profiles of the rolls are designed to produce the desired geometry.

3. Powder Compaction

Powder compaction can be characterized as mass conserving, involving the granular state of material, and a mechanical basic process of flow and plastic deformation. In this context, only compaction of metal powders is mentioned, but generally, compaction of molding sand, ceramic materials, etc., can also belong to this category.

In the compaction of metal powders, as shown in Figure 15.1 (c), the die cavity is filled with a measured volume of powder and compacted at pressures typically around 500 N/mm^2. During this pressing phase, the particles are packed together and plastically deformed. Typical densities after compaction are 80% of the density of the solid material. Because of the plastic deformation, the particles are "welded" together, giving sufficient strength to withstand handling. After compaction, the components are heat-treated and sintered, normally at 70%~80% of the melting temperature of the material. The sintering atmosphere must be controlled to prevent oxidation. The duration of the sintering process varies between 30min and 2h. The strength of the components after sintering can, depending on the material and process parameters, closely approach the strength of the corresponding solid material.

The die cavity, in the closed position, corresponds to the desired geometry. Compaction machinery includes both mechanical and hydraulic presses. Production rates vary between 6 and 100 components per minute.

4. Casting

【拓展视频】

【拓展视频】

Casting can be characterized as mass conserving, involving the fluid state of material, and a mechanical basic process of filling the die cavity. Casting is one of the oldest manufacturing methods and one of the best-known processes. The material is melted and poured into a die cavity corresponding to the desired geometry, as shown in Figure 15.1 (d). The fluid material takes the shape of the die cavity, and this geometry is finally stabilized by the solidification of the material.

【拓展视频】

【拓展视频】

(a) Drop Forging

(b) The Rolling of plates or Sheets

(c) The Compaction of Metal Powders

(d) Casting

Figure 15.1　Mass-conserving Processes in the Solid State of the Work Material

The stages or steps in a casting process are making a suitable mold, melting the material, pouring the material into the cavity, and solidification. Depending on the mold material, different properties and dimensional accuracies are obtained. Equipment used in a casting process includes furnaces, mold-making machinery, and casting machines.

5. Turning

Turning can be characterized as mass reducing, involving the solid state of work material, and a mechanical primary basic process of fracture. The turning process, the best-known and the most widely used mass-reducing process, is employed to manufacture all types of cylindrical shapes by removing material in the form of chips from the work material with a cutting tool, as shown in Figure 15.2 (a). The work material rotates, and the cutting tool is fed longitudinally. The cutting tool is much harder and more wear-resistant than the work material. A variety of types of lathes are employed, some of which are automatic in operation. The lathes are usually powered by electric motors which, through various gears, supply the necessary torque to the work material and provide the feed motion to the tool.

A wide variety of machining operations or processes based on the same metal-cutting principle are available; among the most common are drilling and milling carried out on various machine tools. By varying the tool shape and the pattern of relative work-tool motions, many different shapes can be produced, as shown in Figure 15.2 (b) and Figure 15.2 (c).

6. Electrical Discharge Machining

Electrical discharge machining (EDM) can be characterized as mass reducing, involving the solid state of work material, and a thermal primary basic process of melting and evaporation, as shown in Figure 15.2 (d). In EDM, material is removed by the erosive action of numerous small electrical discharges (sparks) between the work material and the tool (electrode), the latter having the inverse shape of the desired geometry. [3] Each discharge occurs when the potential difference between the work material and the tool is large enough to cause a breakdown in the fluid medium, fed into the gap between the tool and workpiece under pressure, producing a conductive spark channel. The fluid medium, normally mineral oil or kerosene, serves as a dielectric fluid and coolant, maintains uniform resistance to current flow, and removes the eroded material. The sparking occurs at rate of thousands of times per second, always at the point where the gap between the tool and workpiece is smallest, developing so much heat that a small amount of material is evaporated and dispersed into the fluid. The material surface has a characteristic appearance composed of numerous small craters.

7. Electrochemical Machining

Electrochemical machining (ECM) can be characterized as mass reducing, involving the solid state of work material, and a chemical primary basic process of electrolytic dissolution, as shown in Figure 15.2 (e). Electrolytic dissolution of the workpiece is established through an electric circuit, where the work material serves as the anode and the tool which is approximately the inverse shape of the desired geometry serves as the cathode. The electrolytes normally used are water-based saline solutions (sodium chloride and sodium nitrate in 10%~30% solutions). The voltage, usually in the range of 5~20V, maintains high current densities, 0.5~2A/mm^2, giving a relatively high removal rate, 0.5~6cm^3/(min·1000A), depending on the work material.

8. Flame Cutting

Flame cutting can be characterized as mass reducing, involving the solid state of work material, and a chemical primary basic process of combustion, as shown in Figure 15.2(f). In flame cutting, the material (a ferrous metal) is heated to a temperature where combustion by the oxygen supply can start.

Theoretically, the heat liberated should be sufficient to maintain the reaction once started, but because of heat losses to the atmosphere and the material, a certain amount of heat must be supplied continuously. A torch is designed to provide heat both for starting and maintaining the reaction. The most widely used is the oxyacetylene cutting torch, where heat is created by the combustion of acetylene and oxygen. The oxygen for cutting is normally supplied through a center hole in the tip of the torch.

The flame cutting process can only by used for easily combustible materials. For other materials, cutting processes based on the thermal basic process of melting have been developed (e.g., plasma cutting). This is why cutting under both thermal and chemical basic processes is utilized.

9. Fine Blanking

Fine blanking is a technique used for production of blanks that are perfectly flat with a cut edge comparable to a machined finish. This quick and easy process is worthy of consideration when the number of parts justifies the cost of a blanking tool, especially given that operations such as shaving are eliminated.

One of the fine blanking methods involves the punch having a round edge and a small clearance. This is best used for blanks but appears to give less satisfactory results when producing holes. In this method, the radius on the edge of a die is selected according to the type, hardness, and thickness of a particular material, coupled with the shape of a profile on the component. The minimum radius that will impart a good result on a component is an essential feature, and this radius can vary from 0.3 to 2mm according to conditions.

The question of punch and die clearances is vital with this tool design, and they are always much closer than those used for conventional blanking tools. As a general guide, a total clearance of 0.01 to 0.03mm will yield good results, and it is emphasized that these are total clearances, not each side of a hole or blank.

Material Forming Processes Lesson 15

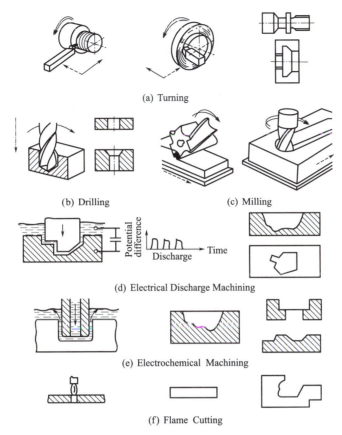

Figure 15.2 Mass-reducing Processes in the Solid State of the Work Material

15.2 Words and Phrases

squeeze	v. 压榨，挤，挤榨
ingot	n. [冶] 锭铁，工业纯铁
profile	n. 剖面，侧面，外形，轮廓
ceramic	adj. 陶器的
sintering	n. 烧结
oxidation	n. [化] 氧化
solidification	n. 凝固
longitudinal	adj. 经度的，纵向的
erosive	adj. 侵蚀性的，腐蚀性的
dissolution	n. 分解，解散
electrolyte	n. 电解，电解液
sodium chloride	氯化钠
sodium nitrate	硝酸钠

ferrous	*adj.* 铁的，含铁的，[化] 亚铁的
oxyacetylene	*adj.* [化] 氧乙炔的
acetylene	*n.* [化] 乙炔，电石气
combustible	*adj.* 易燃的
plasma	*n.* [解] 血浆，乳浆，[物] 等离子体，等离子区
fine blanking	精密冲裁
cut edge	剪切刃
clearance	*n.* 间隙，（公差中的）公隙
shaving	*n.* 刨削，修整
punch	*n.* 冲头，凸模，冲压机，冲床，打孔机；*vt.* 冲孔，打孔
blank	*n.* 落料件
producing hole	冲孔
die	*n.* 模具，凹模

15.3 Complex Sentence Analysis

[1] Technically, forging is defined as the process of enhancing metal utility by shaping, refining, and improving its mechanical properties through controlled plastic deformation under impact or pressure.

① be defined as…：被定义为……

② under impact or pressure：在冲击力或压力的作用下。

[2] Excess material is squeezed out between the die faces at the periphery as flash, which is removed in a later trimming process.

① which 引导定语从句，修饰 excess material。

② flash：飞边。

③ trimming process：切边过程，清理过程。

[3] Each discharge occurs when the potential difference between the work material and the tool is large enough to cause a breakdown in the fluid medium, fed into the gap between the tool and workpiece under pressure, producing a conductive spark channel.

① when 引导时间状语从句，修饰 occurs。

② fed into…修饰 fluid medium。

15.4 Exercise

Translate the following paragraph

Metal forming is the direct alteration of form, surface, and material properties of a

Material Forming Processes Lesson 15

workpiece while preserving mass and cohesion. The processes of forming use the plasticity of metals for the production of semi-finished materials and structural parts. Forming is based on the flexible moldability of numerous materials and relies on the ability of the material structure to bear unit deformations along crystalline gliding planes without damaging the material cohesion. Forming saves materials as no waste is produced. The objective is to obtain a finished surface to avoid expensive finishing. Metal forming processes are classified according to the effective stresses into forming under compressive, a combination of tensile and compressive, tensile, bending, and shearing conditions. Important processes are upsetting, wire drawing, deep drawing, extruding, stretch forming, bending, and forging. The forming process can be influenced by the workpiece, the tool, the lubricant, the environment medium, and the machine. It can be describedthrough yield stress, deformation size, flow conditions, anisotropy, and the flow curve. Processing heated workpieces is called hot-working; processing workpieces at ambient temperature is called cold-working. With metallic materials cold forming is usually accompanied by work hardening, which is characterized by the fact that with increasing deformation, yield strength and breaking strength increase, and breaking elongation decreases.

Lesson 16　Introduction to Mould

16.1　Text

Moulds are fundamental technological devices for industrial production. Industrially produced goods are formed in moulds designed and built specially for them. The mould is the core part of manufacturing process because its cavity gives the product its shape. There are many kinds of moulds, such as casting and forging dies, ceramic moulds, die-casting moulds, drawing dies, injection moulds, glass moulds, magnetic moulds, metal extruding moulds, plastic and rubber moulds, plastic extruding moulds, powder metallurgical moulds, compressing moulds, etc.

The following is an introduction to some moulding processes and the corresponding moulds used.

1. Compression Moulding

【拓展视频】

【拓展视频】

Compression moulding is the least complex of the moulding processes and is ideal for large parts or low-quantity production. For low-quantity requirements, it is more economical to build a compression mould than an injection mould. Compression moulds are often used for prototyping, where samples are needed for testing fit and forming into assemblies. This allows for further design modification before building an injection mould for high-quantity production. Compression moulding is the best suited for designs where tight tolerances are not required.

A compression mould consists of simply two plates with cavities cut into either one or both plates. Additional plates between the top and bottom plates can create cavities in the moulded part. Figure 16.1 shows a basic two-plate single cavity mould. The mould does not require heater elements or temperature controllers. The moulding temperature is fully controlled by the pressure it operates under.

Due to the simplicity of the mould, it is the most economical mould to purchase, keeping small quantity runs affordable.

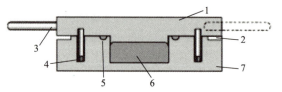

1—top plate; 2—opening bar slot; 3—handle;
4—dowel pin & bushing; 5—flash & tear trim gate;
6—part cavity; 7—bottom plate.

Figure 16.1　A Basic Two-plate Single Cavity Mould

2. Injection Moulding

Injection moulding is the most complex of the moulding processes. Due to the more complex design of the injection mould, it is more expensive to purchase than a cast or compression mould. [1]Although tooling costs can be high, cycle time is much shorter than other processes, and the part cost can be low, particularly when the process is automated. [2]Injection moulding is well-suited for moulding delicately shaped parts because high pressure (up to 29,000 psi) is maintained on the material to push it into every corner of the mould cavity.

Moulds used in injection moulding consist of two halves: one stationary half and one movable half. The stationary half is fastened directly to the stationary platen and is in direct contact with the nozzle of the injection unit during operation. The movable half is secured to the movable platen and usually contains the ejector mechanism. A balanced runner system carries the plastic from the sprue to each individual cavity.

An injection mould can be a simpler two-plate mould with a runner system to allow the rubber compound to be injected into each cavity from the parting line or a more complex mould with a number of plates, an ejector system, and additional heating elements within the core.

Figure 16.2 shows basic three-plate and two-plate multi-cavity injection moulds. The moulds do not require heater elements or temperature controllers. The moulding temperature is controlled by the injection pressure it operates under.

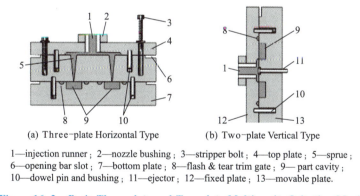

(a) Three-plate Horizontal Type   (b) Two-plate Vertical Type

1—injection runner; 2—nozzle bushing; 3—stripper bolt; 4—top plate; 5—sprue; 6—opening bar slot; 7—bottom plate; 8—flash & tear trim gate; 9—part cavity; 10—dowel pin and bushing; 11—ejector; 12—fixed plate; 13—movable plate.

Figure 16.2 Basic Three-plate and Two-plate Multi-cavity Injection Moulds

3. Cast Moulding

There are two types of casting: open casting and pressure casting. In open casting, the liquid mixture is poured into the open cavity, the cap is put in place, and the cavity is pressurized. Pressure casting is used for more complex parts and when moulding foam materials.

In principle, pressure casting is identical to injection moulding but with a different class of materials. Cast moulding can produce parts with identical geometries to injection-moulded ones. In many cases, injection moulding has been a substitute for casting moulding due to decreased part costs. However, for structural parts, particularly those with thick walls, cast moulding can often be the better selection.

Because the materials flow as low-pressure liquids, tooling is generally less expensive. Low tooling costs make casting ideal for small production quantities and prototypes. It is suited for short to medium-production runs. Liquid cast urethane compounds have outstanding resistance to abrasion, impact, and flex fatigue. Also, complex shapes and thick cross-sections can be produced in many compounds. However, this process has longer cycle and cure times than other moulding processes. Once the material has been moulded, it cannot be reground and reused.

The construction of the mould for cast moulding is almost identical to that of moulds for injection moulding. It consists of two major sections—the ejector half and the cover half, which meet at the parting line. The cavities and cores are usually machined into inserts that are fitted into each of these halves. The cover half is secured to the stationary platen, while the ejector half is fastened to the movable platen. The cavity and machining core must be designed so that the mould halves can be pulled away from the solidified casting. Ejector pins are required to remove the part from the mould when it opens. Lubricants are also sprayed into the cavities to prevent sticking.

Moulds are usually made of tool steel, mould steel, or maraging steel. [3] Since the mould materials have no natural porosity and the molten metal rapidly flows into the mould during injection, venting holes and passageways must be built into the moulds at the parting line to evacuate the air and gases in the cavity.

4. Extrusion Moulding

[4] Although extrusion moulds are quite simple, the extrusion moulding process requires great care in setting up and processing to ensure product consistency. Pressure is forced through the die plate that has the correct profile cut into it. Variations in feed rate, temperature, and pressure need to be controlled.

Unlike compression or injection moulding, the rubber is not cured when it comes out of the mould. The raw rubber is laid out on circular or long trays (depending on the profile) and loaded into an autoclave for curing under heat and pressure.

For long continuous lengths, a salt bath curing system may be used, and for silicone-extrusion, a continuous heating line is used. The curing process used is dependent on the quantity and profile of the extrusion required.

Most extrusion moulds are simply cylindrical steel with the profile of the intended extrusion wire cut into them. Allowances are made for the shrinkage and expansion of the intended compound. Extrusion dies are the least complex of the moulds.

These moulds are relatively cheap to build, but because of the processing involved, minimum run quantities will vary.

16.2　Words and Phrases

mould	n. 模具，模（型），模塑，压模
cavity	n. 模腔，型腔，空洞
casting die	铸模
forging die	锻模
ceramic mould	陶瓷模
die-casting mould	压铸模
drawing die	拉丝模
injection mould	注塑模
magnetic mould	磁铁成形模
extruding mould	挤压成形模
powder metallurgical mould	粉末冶金模
compressing mould	冲压模
moulding	n. 成形
prototyping	样模制作
tolerance	n. 公差，准许
affordable	adj. 提供得起的
fasten	v. 连接，固定，夹紧
secure	v. 固定，紧固
ejector	n. 脱模销，推顶器
runner	n. 浇道，流道
sprue	n. 流道，浇道
parting line	合模线，拼缝线
core	n. 模芯，中间层
cure	v. 固化，塑化
urethane	n. 聚氨酯
abrasion	n. 磨蚀，磨损
flex fatigue	弯曲疲劳
insert	n. 插件，嵌件
pin	n. 销，杆
tool steel	工具钢
maraging steel	马氏体时效钢
porosity	n. 多孔性，有孔性
venting hole	排气孔，通风孔
evacuate	v. 排出，抽空
autoclave	n. 高压釜
salt bath curing system	盐浴固化系统

silicone extrusion　　　　　　　　硅橡胶挤压

16.3　Complex Sentence Analysis

［1］ Although tooling costs can be high, cycle time is much faster than other processes, and the part cost can be low, particularly when the process is automated.
① tooling costs：模具成本。
② when 引导时间状语从句。

［2］ Injection moulding is well-suited for moulding delicately shaped parts because high pressure (up to 29,000 psi) is maintained on the material to push it into every corner of the mould cavity.
① is well suited for：适用于。
② delicately shaped parts：外形精致的零件。

［3］ Since the mould materials have no natural porosity and the molten metal rapidly flows into the mould during injection, venting holes and passageways must be built into the moulds at the parting line to evacuate the air and gases in the cavity.
since 引导原因状语从句，意为"由于……"。

［4］ Although extrusion moulds are quite simple, the extrusion moulding process requires great care in setting up and processing to ensure product consistency.
to ensure product consistency 为动词不定式作目的状语，以确保产品设计与制造一致。

16.4　Exercise

Translate the following paragraphs

The most common types of moulds used in manufacturing industry today are two-plate moulds, three-plate moulds, side-action moulds, and unscrewing moulds.

A two-plate mould consists of two active plates, into which the cavity and core inserts are mounted. In this mould type, the runner system, sprue, runners, and gates solidify with the part being moulded are ejected as a single connected item. Therefore, the operation of a two-plate mould usually requires continuous machine attendance.

The three-plate mould consists of the stationary or runner plate which contains the sprue and half of the runner, the middle or cavity plate which contains the other half of the runner, the gates, and the cavities. The cavity plate is allowed to float when the mould is open. The movable or core plate contains the cores and the ejector system. This type of mould design facilitates the separation of the runner system and the part when the mould opens.

Lesson 17　Mould Design and Manufacturing

17.1　Text

CAD and CAM are widely applied in mould design and mould making. [1]CAD allows you to draw a model on screen, view it from every angle using 3D animation, and test it by introducing various parameters (pressure, temperature, impact, etc.) into digital simulation models. CAM allows you to control manufacturing quality. The advantages of these computer technologies are numerous: shorter design times (modifications can be made quickly), lower cost, faster manufacturing, etc. This new approach also allows for shorter production runs and last-minute changes to the mould for a particular part. Additionally, these new processes can be used to make complex parts.

1. Computer Aided Design (CAD) of Mould

Traditionally, creating drawings of mould tools has been a time-consuming task that is not part of the creative process. Drawings are an organizational necessity rather than a desired part of the process.

CAD simplifies and enhances the design process using computers and peripheral devices. CAD systems offer an efficient means of design and can be used to create inspection programs alongside coordinate measuring machines and other inspection equipment. CAD data can also play a critical role in selecting process sequence.

A CAD system consists of three basic components: the hardware, the software, and the user. The hardware components typically include a processor, a system display, a keyboard, a digitizer, and a plotter. The software component consists of programs which allow it to perform design and drafting functions. The user is the tool designer who uses the hardware and software to perform the design process.

Based on the 3D data of the product, the core and cavity have to be designed first. Usually, the designer begins with a preliminary part design, which means the work around the core and cavity could change. Modern CAD systems can support this by calculating a split line for a defined draft direction, splitting the part into the core and cavity side, and generating run-off or shut-off surfaces. After calculating the optimal draft of the part, the position and direction of the cavity, slides, and inserts have to be defined. In the conceptual stage, the positions and geometry of mould components—such as slides and the ejection system—are roughly defined. With this information, the size and thickness of the plates can be defined, and the corresponding standard mould can be chosen from the standard catalog. If no standard moulds fit these needs, the closed standard mould is chosen and modified accordingly—by adjusting the constraints and parameters so that any number of plates with any size can be used in the mould. Detailing the functional components and adding the standard components complete the mould. This all happens in 3D solid model of mould (Figure 17.1). Moreover, the mould system provides functions for checking, modifying, and detailing the part. In this early stage, drawings and bills of materials can be created automatically.

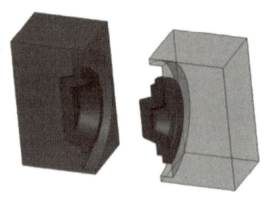

Figure 17.1 3D Solid Model of Mould

Through the use of 3D and the intelligence of the mould design system, typical 2D mistakes—such as collisions between cooling components and cavities or wrong hole positions—can be eliminated at the beginning. [2] At any stage, a bill of materials and drawings can be created, allowing material to be ordered on time and maintaining an up-to-date document for discussions with the customer or bids for a mould base manufacturer.

A special 3D mould design system can shorten development cycles, improve mould quality, enhance teamwork, and free the designer from tedious routine work. Economical success, however, is highly dependent on the organization of the workflow. Development cycles can be shortened only when organizational and personnel measures are taken. The part design, mould design, electric design, and mould manufacturing departments have to consistently work together in a tight relationship.

2. Computer-aided Manufacturing (CAM) of Mould

One way to reduce manufacturing costs and lead-time is by setting up a manufacturing system that maximizes the use of equipment and personnel. The foundation for this type of manufacturing system is the use of CAD data to aid in making key process decisions that ultimately improve machining precision and reduce non-productive time. This is called computer-aided manufacturing (CAM). [3] The objective of CAM is to produce, if possible, sections of a mould without intermediate steps by initiating machining operations from the computer workstation.

With a good CAM system, automation does not just occur within individual features. Automation of machining processes also occurs between all features that make up a part, resulting in tool-path optimization. As you create features, the CAM system constructs a process plan. Operations are ordered based on a system analysis to reduce tool changes and the number of tools used.

On the CAM side, the trend is toward newer technologies and processes such as micro milling to support the manufacturing of high-precision injection moulds with complex 3D structures and high surface qualities. CAM system will continue to enhance machining intelligence until the CNC programming process becomes completely automatic. This is especially true for advanced multifunction machine tools that require flexible combinations of machining operations. CAM system will increasingly automate redundant manufacturing tasks that computers can handle faster and more accurately, while retaining necessary machinist's control.

[4] With the emphasis in the mould-making industry on producing moulds efficiently while maintaining quality, mould makers need to keep up with the latest software technologies that allow them to program and cut complex moulds quickly, reducing production time. The industry is moving toward improving data exchange quality between CAD and CAM, as well as CAM and CNC, with CAM system becoming more "intelligent" regarding machining processes—resulting in reduced cycle time and overall machining time. Five-axis machining is also emerging as essential on the shop floor, especially for deep cavities. With the introduction of electronic data processing (EDP) into the mould-making industry, new opportunities have arisen to shorten production time, improve cost efficiencies, and achieve higher quality.

17.2　Words and Phrases

screen	n. 屏幕，隔板
animation	n. 动画
digital simulation model	数字模拟模型
legion	n. 多，大批，无数
creative	adj. 有创造力的
peripheral	adj. 外围的，周边的
split	adj. 分割的，对分的
run-off	n. 流出口，流放口
shut-off	n. 截流，断流
ejection system	脱模系统，卸料装置
collision	n. 打击，碰撞

tedious	*adj.* 沉闷的
lead-time	*n.* 研制周期
tool-path	*n.* 方法路径
multifunction	*n.* 多功能
redundant	*adj.* 多余的，冗余的

17.3　Complex Sentence Analysis

[1] CAD allows you to draw a model on screen, view it from every angle using 3D animation, and test it by introducing various parameters (pressure, temperature, impact, etc.) into digital simulation models.

　　draw、view 和 test 表示三个并列的操作。

[2] At any stage, a bill of materials and drawings can be created, allowing the material to be ordered on time and maintaining an up-to-date document for discussions with the customer or bids for a mould base manufacturer.

① a bill of materials：材料清单。

② allowing 和 maintaining 引导两个现在分词短语，作伴随状语。

[3] The objective of CAM is to produce, if possible, sections of a mould without intermediate steps by initiating machining operations from the computer workstation.

① if possible：插入语，如果可能，是 if it is possible 的省略形式。

　　试比较：

　　a. If possible, I will visit you in Chicago next month.

　　b. I will lend you some money to help you with the present difficulty if possible.

② by…：通过……方式。

[4] With the emphasis in the mould-making industry on producing moulds efficiently while maintaining quality, mould makers need to keep up with the latest software technologies that allow them to program and cut complex moulds quickly, reducing production time.

① with…：随着……，引导介词短语作状语，修饰整个句子。

　　试比较：

　　With the development of the economy in China, the people around the country are living a happy life.

② efficiently 修饰 producing moulds。

③ keep up with：紧跟。

④ that 引导宾语从句，修饰 software technologies。

17.4　Exercise

Translate the following paragraphs

A key decision early in the mould-making process is determining which machining operations will be used and in what order. Machining considerations should be analyzed during the development of the CAD model. If this hasn't done, the programmer may not be able to use certain machining strategies.

Each process has advantages and disadvantages when producing a close tolerance mould. Proper selection of process and sequence will not only result in more precise dimensional control but also reduce manufacturing time by minimizing bench work.

In the worst case, the model may have to be modified, significantly adding to lead-time. Not all workpieces are suitable for hard milling. The smallest internal radius, the largest working depth, and the hardness of the material all have to be considered when making this decision.

CAD data can be used to program electrode machining operations. Dimensional data can be downloaded to software that automates electrode design and generates simulations of electrode action, allowing users to test cuts prior to burning. The software also enables users to try different qualities of graphite to determine the optimum grade before actual burning begins.

Lesson 18　Heat Treatment of Metal

18.1　Text

The generally accepted definition for heat treating metals and metal alloys is "heating and cooling a solid metal or alloy to obtain specific conditions and/or properties". Heating solely for hot-working (as in forging operations) is excluded from this definition. Likewise, heat treatment, used for products such as glass or plastics are also excluded.

1. Isothermal Transformation Diagram

The basis for heat treatment is the time-temperature-transformation diagram or TTT diagram, where all three parameters are plotted in a single diagram.

To plot TTT diagram, a particular steel is held at a given temperature, and the structure is examined at predetermined intervals to record the amount of transformation. [1] It is known that eutectoid steel (T8) under equilibrium conditions contains all austenite above 727℃, whereas below, it is pearlite. To form pearlite, carbon atoms must diffuse to form cementite. This diffusion, being a rate process, requires sufficient time for the complete transformation of austenite to pearlite. From different samples, it is possible to note the amount of transformation at any temperature. [2] These points are then plotted on a graph with time and temperature as the axes. Transformation curves can be plotted through these points, as shown in Figure 18.1 for eutectoid steel (T8). The curve on the extreme left represents the time required for the transformation of austenite to pearlite to start at any temperature. Similarly, the curve on the extreme right represents the time required for completing the transformation. Between the two curves are points representing partial transformation. The horizontal lines M_s and M_f represent the start and finish of martensitic transformation.

【拓展视频】

2. Classification of Heat Treating Processes

In some instances, heat treatment procedures are clear-cut in technique and application, whereas in others, simple explanations are insufficient because the same technique may be used for different objectives. For example, stress relieving and tempering are often accomplished with the same equipment and identical time and temperature cycles. The objectives, however, are different for the two processes.

The following descriptions of the principal heat treating processes are generally arranged according to their interrelationships.

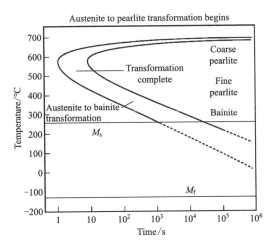

Figure 18.1 Isothermal Transformation Diagram of Eutectoid Steel (T8)

(1) Normalizing.

Normalizing consists of heating a ferrous alloy to a suitable temperature above its specific upper transformation temperature. [3] This is followed by cooling in still air to at least some temperature well below its transformation temperature range. For low-carbon steels, the resulting structure and properties are the same as those achieved by full annealing; for most ferrous alloys, normalizing and annealing are not synonymous.

Normalizing is usually used as a conditioning treatment for refining the grains of steels subjected to high temperatures for forging or other hot-working operations. The normalizing process is usually succeeded by another heat treating operation such as austenitizing for hardening, annealing, or tempering.

(2) Annealing.

Annealing is a generic term denoting a heat treatment that consists of heating to and holding at a suitable temperature, followed by cooling at a suitable rate. It is primarily used to soften metallic materials but also to produce desired changes in other properties or microstructures. The purpose may include improving machinability, facilitating cold-working (known as in-process annealing), improving mechanical or electrical properties, or increasing dimensional stability. When applied solely to relieve stresses, it is commonly called stress-relief annealing, synonymous with stress relieving.

【拓展视频】
【拓展视频】

When "annealing" is applied to ferrous alloys without qualification, full annealing is implied. This is achieved by heating above the alloy's transformation temperature, then applying a cooling cycle which provides maximum softness. This cycle may vary widely, depending on compositions and characteristics of the specific alloy.

(3) Quenching.

Quenching is the rapid cooling of a steel or alloy from the austenitizing temperature by immersing the workpiece in a liquid or gaseous medium. Quenching media commonly include water, 5% brine, 5% caustic in an aqueous solution, oil, polymer solutions, or gas (usually air or nitrogen).

Selection of a quenching medium depends largely on the hardenability and mass (principally section thickness) of the material being treated.

The cooling capabilities of the above-listed quenching media vary greatly. In selecting a quenching medium, it is a good practice to avoid a solution with more cooling power than needed to minimize the possibility of cracking and warping of the parts being treated. Modifications of quenching include direct quenching, fog quenching, hot quenching, interrupted quenching, selective quenching, spray quenching, and time quenching.

(4) Tempering.

In heat treating ferrous alloys, tempering involves reheating the austenitized and quench-hardened steel or iron to some preselected temperature below the lower transformation temperature (generally below 704℃). Tempering offers a means of obtaining various combinations of mechanical properties. Tempering temperatures for hardened steels are often no higher than 150—200℃. The term "tempering" should not be confused with either process annealing or stress relieving. Although time and temperature cycles for the three processes may be the same, the conditions of the materials being processed and the objectives may differ.

(5) Stress Relieving.

Like tempering, stress relieving is always done by heating to some temperature below the lower transformation temperature for steels and irons. For nonferrous metals, the temperature may vary from slightly above room temperature to several hundred celsius degrees, depending on the alloy and the amount of stress relief desired.

The primary purpose of stress relieving is to relieve stresses imparted to the workpiece from processes like forming, rolling, machining, or welding. The usual procedure is to heat workpieces to the pre-established temperature long enough to reduce the residual stresses (a time and temperature-dependent operation) to an acceptable level, followed by cooling at a relatively slow rate to avoid creatiing new stresses.

18.2　Words and Phrases

forge	v. 锻造
eutectoid	adj. 共析的
austenite	n. 奥氏体
pearlite	n. 珠光体
martensitic	adj. 马氏体的

stress relieving		消除应力,低温退火
tempering		*n*. 回火
normalizing		*n*. 常化,正火
ferrous alloy		铁合金
transformation		*n*. 变换,转换,相变
still		*adj*. 不动的,静止的
full annealing		完全退火
notably		*adv*. 显著地,特别地
austenitize		*v*. 奥氏体化,使成奥氏体
denote		*v*. 指示,表示,概述
machinability		*n*. 切削加工性,机械加工性能
facilitation		*n*. 便于
in-process		*adj*. (加工、处理)过程中的
qualification		*n*. 资格,条件,限制,限定
quenching		*n*. 淬火
brine		*n*. 盐水
caustic		*adj*. 腐蚀性的,碱性的
aqueous		*adj*. 水的,水成的
warp		*n*. 翘曲,变形
glossary		*n*. 词汇表,术语汇编
quench-hardened		*adj*. 淬火硬化的
process annealing		工序间退火,中间退火
fog quenching		喷雾淬火
hot quenching		高温淬火,热淬火
interrupted quenching		分级淬火
selective quenching		局部淬火
time quenching		等温淬火,控制时间淬火,即时淬火

18.3　Complex Sentence Analysis

[1] It is known that eutectoid steel (T8) under equilibrium conditions contains all austenite above 727℃, whereas below, it is pearlite.

共析钢(T8)在平衡条件下,在727℃以上时全为奥氏体,低于此温度则为珠光体。

① that 引导主语从句。
② eutectoid steel:共析钢。
③ below 在这里指低于727℃。

[2] These points are then plotted on a graph with time and temperature as the axes.

with time and temperature as the axes：axes 在此处指坐标轴，以时间和温度为坐标轴。

[3] This is followed by cooling in still air to at least some temperature well below its transformation temperature range.

① follow：接着，跟着。

② this 指上一句内容，先 heating a ferrous alloy…，再 cooling…

18.4 Exercise

Translate the following paragraphs

Carburizing involves the absorption and diffusion of carbon into solid ferrous alloys by heating above the upper transformation temperature of the specific alloy. Temperatures used for carburizing generally range from 900℃ to 1040℃. Heating is done in a carbonaceous environment (liquid, solid, or gas). This produces a carbon gradient extending inward from the surface, enabling the surface layers to be hardened to a high degree either by quenching from the carburizing temperature or by cooling to room temperature followed by reaustenitizing and quenching.

Carbonitriding is a surface hardening process in which a ferrous material (most often a low-carbon grade of steel) is heated above the transformation temperature in a gaseous atmosphere, causing simultaneous absorption of carbon and nitrogen at the surface. By diffusion, a concentration gradient is created. The process is completed by cooling at a rate that produces the desired properties in the workplace.

Lesson 19 Numerical Control System

19.1 Text

[1] Numerical control (NC) system is a form of programmable automation in which processing equipment is controlled by numbers, letters, and symbols. These are coded to define a program of instructions for a particular workpiece or job. The instructions are provided by either the Electronic Industries Association (EIA) code or the American Standard Code for Information Interchange (ASCII). ASCII-coded machine control units will not accept EIA-coded instructions and vice versa. Increasingly, however, control units accept instructions in either code. [2] Automation by NC is adaptable to all metalworking machines. Lathes, milling machines, drill presses, boring machines, grinding machines, turret punches, flame or wire-cutting and welding machines, and even pipe benders are available with numerical controls.

1. Basic Components of Numerical Control System

A numerical control system consists of three basic components:

(1) Program instructions.

The program instructions are the detailed step-by-step commands that direct the processing equipment. [3] In its most common form, the commands refer to positions of a machine tool spindle with respect to the worktable on which the workpiece is fixed. More advanced instructions include selection of spindle speeds, cutting tools, and other functions.

(2) Machine control unit.

The machine control unit (MCU) consists of the electronics and control hardware that read and interpret the program of instructions and convert it into mechanical actions of the machine tool or other processing equipment.

(3) Processing equipment.

The processing equipment is the component that performs metal process. In the most common example of numerical control, it is used to perform machining operations. The processing equipment consists of the worktable and spindle, as well as the motors and controls needed to drive them.

2. Types of Numerical Control Systems

There are two basic types of NC systems: point-to-point NC system and contouring

NC system.

Point-to-point NC systems, also called positioning NC systems, are simpler than contouring NC systems. Their primary purpose is to move a tool or workpiece from one programmed point to another. Usually, the machine function, such as a drilling operation, is also activated at each point by command from the NC Program. Point-to-point NC systems are suitable for hole machining operations such as drilling, countersinking, counterboring, reaming, boring, and tapping. Hole punching machines, spot welding machines, and assembly machines also use point-to-point NC systems.

Contouring NC systems, also known as continuous path systems, involve positioning and cutting operations both along controlled paths but at different velocities. Because the tool cuts as it travels along a prescribed path, accurate control and synchronization of velocities and movements are important. The contouring NC systems are used on lathes, milling machines, grinders, welding machinery, and machining centers. Movement along the path, or interpolation, occurs incrementally, by one of several basic methods. Several interpolation schemes have been developed to deal with various challenges encountered in generating a smooth continuous path with a contouring NC system. They include linear interpolation, circular interpolation, helical interpolation, parabolic interpolation, and cubic interpolation. In all interpolations, the path controlled is that of the center of rotation of the tool. Compensation for different tools, different diameter tools, or tools wear during machining can be made in the NC program.

3. Programming for Numerical Control

[4] A program for NC consists of a sequence of directions that causes an NC machine to carry out a certain operation, with machining being the most commonly used process. Programming for NC can be done by an internal programming department, on the shop floor, or purchased from an outside source. Additionally, programming for NC may be done manually or with assistance of computer.

The program contains instructions and commands. Geometric instructions pertain to relative movements between the tool and the workpiece. Processing instructions pertain to spindle speeds, feeds, tools, etc. Travel instructions pertain to the type of interpolation and slow or rapid movements of the tool or worktable. Switching commands pertain to the on/off position for coolant supplies, spindle rotation, direction of spindle rotation, tool changes, workpiece feeding, clamping, etc. The first NC programming language was developed by MIT's developmental work on NC programming systems in the late 1950s and was called APT (automatically programmed tools).

4. Distribute Numerical Control and Computer Numerical Control

The development of NC was a significant achievement in batch and job shop manufacturing, from both technological and commercial viewpoints. There have been two enhancements and extensions of NC technology, including:

Numerical Control System Lesson 19

(1) Distribute numerical control (DNC).

DNC can be defined as a manufacturing system in which a number of machines are controlled by a computer through direct connection and in real time. The tape reader is omitted in DNC, thus relieving the system of its least reliable component. Instead of using the tape reader, the part program is transmitted to the machine tool directly from the computer memory. In principle, one computer can be used to control more than 100 separate machines. The DNC computer is designed to provide instructions to each machine on demand. When the machine needs control commands, they are communicated to it immediately.

Since the introduction of DNC, there have been dramatic advances in computer technology. The physical size and the cost of a digital computer has been significantly reduced, while its computational capabilities have been substantially increased. In NC, the result of these advances has been that the large hard-wired MCUs of conventional NC have been replaced by control units based on the digital computer. Initially, minicomputers were utilized in the early 1970s. As further miniaturization occurred in computers, minicomputers were replaced by today's microcomputers.

(2) Computer numerical control (CNC).

CNC is an NC system using a dedicated microcomputer as the MCU. Because a digital computer is used in both CNC and DNC, it is appropriate to distinguish between the two types of systems. There are three principal differences:

(1) DNC computers distribute instructional data to, and collect data from, a large number of machines. CNC computers control only one machine or a small number of machines.

(2) DNC computers occupy a location that is typically remote from the machines under their control. CNC computers are located very near their machine tools.

(3) DNC software is developed not only to control individual pieces of production equipment but also to serve as part of a management information system in the manufacturing sector of the firm. CNC software is developed to augment the capabilities of a particular machine tool.

19.2 Words and Phrases

instruction	*n.* 指令
binary	*adj.* 二进制的；*n.* 二进制
lathe	*n.* 车床
mill	*v.* 铣
drill	*v.* 钻
bore	*v.* 镗

grind	v.	磨
turret	n.	转盘
punch	n.	冲床
flame	n.	（电）火花
wire-cutting		线切割
pipe bender		弯管机
spindle	n.	主轴
contour	n.	轮廓
workpiece	n.	工件
countersink	n.	钻（沉头）孔
counterbore	n.	镗（沉头）孔
ream	n.	铰孔
tapping	n.	攻螺纹
spot welding		点焊
synchronization	n.	同步
interpolation	n.	插补
parabolic	adj.	抛物线的
compensation	n.	补偿
pertain	v.	合适
coolant	n.	冷却液
clamping	n.	夹紧
miniaturization	n.	小型化
dedicated	adj.	专用的

19.3 Complex Sentence Analysis

[1] Numerical control（NC）is a form of programmable automation in which processing equipment is controlled by numbers, letters, and other symbols.
① a form of…：一个……的形式。
② in which 引导定语从句，修饰 programmable automation。
③ by…：通过……，使用……方法。

[2] Automation by NC is adaptable to all metalworking machines. Lathes, milling machines, drill presses, boring machines, grinding machines, turret punches, flame or wire-cutting and welding machines, and even pipe benders are available with numerical controls.
① is adaptable to…：可适用于……
② are available with…：用……有效。

[3] In its most common form, the commands refer to positions of a machine tool

spindle with respect to the worktable on which the part is fixed.

① refer to…：指……
② with respect to…：相对于……
③ on which 引导定语从句，修饰 worktable。

[4] A program for numerical control consists of a sequence of directions that causes an NC machine to carry out a certain operation, machining being the most commonly used process.

① consist of：包含。
② that 引导定语从句，修饰 directions。
③ machining 是名词，意为"机械加工"。
④ being 是分词短语。

19.4 Exercise

Translate the following paragraphs

Numerical control can be defined as a method of accurately managing the movement of machines by a series of programmed numerical data which activate the motors of the machine. There is nothing complex or magical about this system. It is based on the simple fundamental that combines automatic measurement of machine table slide with a series of programmed instructions.

The NC machine is supplied with detailed information regarding the part by means of punched tape. The machine decodes this punched information, and electronic devices activate the various motors on the machine, causing them to position the work and follow specific machining instructions. The measuring and recording devices incorporated into NC machines assure that the part being manufactured will be accurate. NC machines minimize the possibility of human error which existed before their development.

Over recent years, there have been great improvements to NC systems. In many cases, the tape reader on NC machines is bypassed, and the machine receives its instructions directly from a computer. However, students should be made aware of the basics of NC before being exposed to the more recent developments of CNC and DNC.

Lesson 20   Virtual Manufacturing

20.1   Text

1. Definition of Virtual Manufacturing

[1] Virtual manufacturing (VM) is an integrated and synthetic manufacturing environment exercised to enhance all levels of decision-making and control in a manufacturing enterprise. VM can be described as a simulated model of the actual manufacturing setup, which may or may not exist. It holds all the information relating to the process, process control and management, and product-specific data. It is also possible to have part of the manufacturing plant be real and the other part virtual. VM involves the use of computer models and simulations of manufacturing processes to aid in the design and production of manufactured products.

Three different types of VM paradigms that use virtual reality (VR) to provide an integrated environment are as follows:

(1) [2] Design-centered VM: It provides designers with tools to design products that meet design criteria such as design for X.

(2) Production-centered VM: It provides means to develop and analyze alternative production and process plans.

(3) [3] Control-centered VM: It allows the evaluation of product design, production plans, control strategy, and a means to improve all of them through the simulation of the control process.

2. Significance of Virtual Manufacturing

VM aims to provide an integrated environment for a number of isolated manufacturing technologies such as CAD, CAM, and CAPP, allowing multiple users to concurrently carry out all or some of these functions without needing to be physically close to each other. For example, a process planning engineer and a manufacturing engineer can evaluate and provide feedback to a product designer, who may be physically located in another state or country, at the same time as the design is being conceived.

Another important contribution of VM is virtual enterprise (VE). VE was defined as a rapidly configured multi-disciplinary network of small-sized and process-specific firms configured to meet a window of opportunity to design and produce a specific product. Using this technology, a group of people or corporations can pool their expertise and resources and capitalize a market opportunity by sharing information in a VM environment. [4] The principal advantage of this technology is its ability to provide a multimedia environment, enhancing communication at all levels in a product's life cycle.

3. Applications of VM

(1) Steel industry.

Baowu Steel introduced AR (augmented reality) technology and applications to create an "AR intelligent operation and maintenance system", which brought a new way of equipment operation and maintenance work to metallurgical enterprises. The system combines new-generation information technologies such as 5G, cloud computing, edge computing, big data, artificial intelligence, AR, etc. to realize digital information visualization, precise remote collaboration, and efficient process record management of related equipment.

(2) Energy industry.

Chint Group introduced AR technology to create an "AR power distribution operation and maintenance system". The system imports the content configuration of the operation instruction book in a visual way. On-site employees can scan the equipment QR (quick response) code to view the operation instruction content by wearing AR glasses, which greatly improves the level of on-site operation standards and the work efficiency of personnel.

(3) Automation industry.

Sinopec Northeast Petroleum Bureau Co., Ltd integrated AR remote communication and collaboration plug-in HiLeia.PS based on the existing GIS (geographic information system) to create an "AR visualization collaboration management platform". The platform is mainly used in two scenarios: fault diagnosis and emergency maintenance, helping back-end experts and on-site operators to communicate visually from the first perspective.

(4) Heavy machinery industry.

In Haier's Shanghai washing machine interconnected factory, the washing machine inner drum production line using digital twin technology is running. By building a three-dimensional digital model, the real-life scene is reproduced, and a "twin" is created in the virtual world to achieve production improvement and optimization, reduce costs, and improve efficiency.

These cases show the wide applications and significant effects of VM in improving production efficiency, optimizing production processes, reducing production costs, and improving product quality. With the continuous development of computer technology and VR, the application prospects of VM will be broader. Please note that the above information is for reference only, and the specific application effect may vary depending on factors such as industry, enterprise, and technology.

20.2　Words and Phrases

virtual manufacturing	虚拟制造
synthetic	*adj.* 人造的，综合的，假想的
be described as	被说成，被称作
paradigm	*n.* 范例，式样
virtual reality	虚拟现实
iteratively	*adv.* 重复，迭代
concurrently	*adv.* 同时，兼，并行地
virtual enterprise	虚拟企业
configure	*v.* 使成形，使具形体，装配，配置
multi-disciplinary	*adj.* （有关）多种学科的
collision	*n.* 撞，冲突，抵触
ergonomical	*adj.* 人类工程（学）的，人机学的
infrastructure	*n.* 基础（结构），基本设施
caterpillar	*n.* 履带式车（辆），履带式挖土机
bulldozer	*n.* 推土机

20.3　Complex Sentence Analysis

［1］ Virtual manufacturing (VM) is an integrated and synthetic manufacturing environment exercised to enhance all levels of decision-making and control in a manufacturing enterprise.

　　exercised：运用的，使用的；作后置定语，修饰 environment。

［2］ Design-centered VM：It provides designers with tools to design products that meet design criteria such as design for X.

① to design products that meet design criteria such as design for X 为动词不定式短语作定语，修饰 tools。

② provide A with B 同 provide B for A：把 B 提供给 A。

　　例如，Parents provide us with food and clothing.

［3］ Control-centered VM：It allows the evaluation of product design, production plans, control strategy, and a means to improve all of them through the simulation of the control process.

　　means：方法，手段，既可指单数又可指复数；如果作主语，应根据意思决定其谓语动词用单数或复数。

　　试比较：

a. Every means has been tried.

b. All means have been tried.

[4] The principal advantage of this technology is its ability to provide a multimedia environment, enhancing communication at all levels in a product's life cycle.

① to provide a multimedia environment 作定语，修饰 ability。

② enhancing communication at all levels in a product's life cycle 为现在分词短语作目的状语。

20.4　Exercise

Translate the following paragraphs

Kinematic simulation or animation can be an important CAD/CAM capability in certain settings. Through animation, users can design a product or process that involves moving components and analyze its behavior without having to build prototypes and conduct live trial runs.

For example, mechanical linkage designs have historically been difficult because testing their actual behavior required engineers to build models or prototypes. In such cases, engineers frequently resort to building cardboard or wooden models, a time-consuming process. With animation, a computer model can be quickly built and displayed. By watching the computer screen, the engineer can easily see how the linkage performs.

Other useful applications of animation include analyzing the performance of a robot in a cell or automated guided vehicles (AGVs) on the shop floor. By simulating the planned behavior of a robot or an AGV, engineers and manufacturing personnel can identify and correct problems before implementing them in a live setting.

Lesson 21　Industrial Big Data for Decision-making in Intelligent Manufacturing

21.1　Text

In the era of big data, the massive amount of data generated by the manufacturing industry is characterized by ultra-high dimensions. Addressing these ultra-high dimensional data, tapping into their potential value, and developing a data flow model suitable for the new manufacturing environment are challenging problems. Under the background of "Industry 4.0", big data-driven analysis will bring significant benefits to the manufacturing sector with the support of emerging technologies. The data analysis process aims to improve the transparency of decision-making. Decision-making based on big data-driven analysis maximizes the whole manufacturing system's function according to the enterprise's internal structure, making effective use of manufacturing resources to maximize economic benefits.

In the age of interconnected intelligent information and knowledge-driven systems, big data sparks a digital revolution. Solutions utilizing big data analysis and intelligent computing are inreasingly employed to simplify the processing of large amounts of data and reduce cognitive load. Companies are increasingly adopting strong strategies driven by big data to improve competitiveness.[1] Big data-driven technology provides an excellent opportunity for today's manufacturing mode to shift from traditional manufacturing to intelligent manufacturing. In recent years, with the smart development of industrial factories, big data analysis has become the main driving force for enterprises to create industrial value, making industrial production more intelligent. The data collected from various sources have been applied to industrial production research, which has shifted from being based on analytical models to being based on data-driven.

Big data analysis represents a revolutionary leap in traditional data analysis. The key characteristics of big data can be summarized and defined by 5H: high capacity (a large amount of data), high speed (data generated and updated at a fast pace), high diversity (data from various sources in different forms), high accuracy, and high value (tremendous potential value hidden in data). In the era of big data in the manufacturing industry, the unique characteristics of big data systems are real-time, dynamic, and adaptive. Compared to traditional data analysis systems, data managed by a big data platform comes from the physical world or virtual digital world. Owing to the variety of data sources, the efficient processing of data highlights a more promising future. The manufacturing industry is undergoing a revolutionary information and intelligent transformation. It is imperative to provide on-demand communication services to ensure high reliability, robust scalability, and availability of manufacturing systems.

[2] Intelligent manufacturing covers many aspects of the manufacturing field, including technology and the integration of the manufacturing sector with information technology, aiming to convert data acquired across the product lifecycle into manufacturing intelligence to yield positive impacts on all aspects of manufacturing (such as intelligent products, intelligent production, intelligent services, etc.). Big data can create real-time solutions to meet challenges across various sectors. The use of big data-driven methods will impact the quality management of production systems. Mining and analyzing data related to product quality can provide decision support for quality control and assurance in the manufacturing system.

[3] Intelligent manufacturing aims to build a highly integrated collaborative production ecosystem, which can respond to the dynamically changing demands and environmental conditions in the whole value chain in real-time. The core of intelligent manufacturing lies in the interconnected and deeply integrated relationship between the physical and digital worlds. The contemporary manufacturing industry strategically focuses on integrating advanced digital information technology into various application fields. With the emergence of "Industry 4.0", multiple aspects of manufacturing production, value creation processes, and business models have undergone tremendous changes. Manufacturing companies of all sizes worldwide are moving towards incorporating intelligence into their operations. Formulating a rational and efficient digital strategy will steadily enhance their competitive advantages.

With the further development of data storage and analysis technology, big data-driven analysis plays a cruical role in creating value in the manufacturing industry. The decision-making methods of leaders are evolving, relying more on big data analysis rather than experience to generate greater manufacturing value. Big data-driven technology offers promising opportunities for the manufacturing industry and forms the foundation for sustainable manufacturing in the future, promoting the implementation and development of "Industry 4.0".

In today's competitive contexts, companies are not only interested in understanding the technical aspects of big data analytics (BDA), but are increasingly focused on learning how to exploit the knowledge and insight-creation potential of the data they possess, and to effectively use this knowledge within their strategic and operative decision-making and innovation processes. Big data has already been a research hotspot in intelligent manufacturing. Intellingert manutacturing discusses the relationship between big data and co-innovation, using big data as a common analytical perspective and as a concept aggregating different research streams such as open innovation, co-creation, and collaborative innovation. It also explores the role of the internet of things (IOT) and big data in the digital transformation of businesses. Furthermore, it addresses the recognition and challenges of big data and microgrid, and outlines areas in the microgrid such as stability improvement, asset management, renewable energy prediction, and decision-making support. It offers an analysis of relevant standards for manufacturing systems within the digital manufacturing platforms (DMP) cluster to identify those standards that might be applicable for zero defects manufacturing (ZDM) and for further projects or manufacturing platform designs. It emphasizes how reliability can support different types of strategic decisions in the context of "Industry 4.0" and highlights the need for research associating management decisions with "Industry 4.0" technologies. However, there is currently no systematic review concerning the relationship between intelligent decision-making and big data in manufacturing. We aim to fill this gap by pointing out how the effective exploitation of big data analytics capabilities (BDAC) is crucial for implementing successful decision-making in the intelligent manufacturing.

21.2　Words and Phrases

intelligent robotic manufacturing	智能机器人制造
big data	大数据
ultra-high dimension	超高维度
data flow model	数据流模型
data analysis	数据分析
decision-making	决策
intelligent information interconnection	智能信息互联
digital revolution	数字革命
big data analysis	大数据分析
intelligent computing	智能计算
quality management	质量管理
production systems	生产系统
data storage	数据存储
research hotspot	研究热点
internet of things (IOT)	物联网
microgrid	微电网
digital manufacturing platforms (DMP)	数字制造平台
zero defects manufacturing (ZDM)	零缺陷制造
strategic decisions	战略决策
management decisions	管理决策

21.3　Complex Sentence Analysis

[1] Big data-driven technology provides an excellent opportunity for today's

manufacturing mode to shift from traditional manufacturing to intelligent manufacturing.

主句是 Big data-driven technology provides an excellent opportunity for today's manufacturing mode，不定式短语 to shift from traditional manufacturing to intelligent manufacturing 作目的状语，描述了 today's manufacturing mode，它不是定语从句，而用于表明主句中机会的具体内容或目的。

[2] Intelligent manufacturing covers many aspects of the manufacturing field, including technology and the integration of the manufacturing sector with information technology, aiming to convert data acquired across the product lifecycle into manufacturing intelligence to yield positive impacts on all aspects of manufacturing (such as intelligent products, intelligent production, intelligent services, etc.).

主句是 Intelligent manufacturing covers many aspects of the manufacturing field，现在分词短语 aiming to convert data acquired across the product lifecycle into manufacturing intelligence to yield positive impacts on all aspects of manufacturing（such as intelligent products, intelligent production, intelligent services, etc.）不是定语从句，而用于修饰主句，说明了智能制造的目的或意图。

[3] Intelligent manufacturing aims to build a highly integrated collaborative production ecosystem, which can respond to the dynamically changing demands and environmental conditions in the whole value chain in real-time.

主句是 Intelligent manufacturing aims to build a highly integrated collaborative production ecosystem，定语从句是 which can respond to the dynamically changing demands and environmental conditions in the whole value chain in real-time，修饰 a highly integrated collaborative production ecosystem，提供了有关该生态系统功能和特性的详细信息。

21.4 Exercise

Translate the following paragraph

Under the trend of economic globalization, intelligent manufacturing has attracted much attention from academics and industry. Related enabling technologies make the manufacturing industry more intelligent. As one of the key technologies in artificial intelligence, big data-driven analysis improves the market competitiveness of the manufacturing industry by mining the hidden knowledge value and the potential ability of industrial big data and helps enterprise leaders make wise decisions in various complex manufacturing environments. This paper provides a theoretical analysis basis for big data-driven technology to guide decision-making in intelligent manufacturing, fully demonstrating the practicability of big data-driven technology in the intelligent manufacturing industry, including key advantages and internal motivation. A conceptual framework of intelligent decision-making based on industrial big data-driven technology is proposed in this study, which provides valuable insights and thoughts for the severe challenges and future research directions in this field.

Lesson 22　Robotics and Computer-integrated Manufacturing

22.1　Text

1. Smart Robotic Manufacturing

Robotics is the science of perceiving and acting in the physical world with computer-controlled mechanical devices. Robotic devices vary from industrial robots, mobile robots, medical and surgery robots, rehabilitation robots, drones, and service robots. Among industrial robots, robotic manipulators are the most common devices. Industrial robots have been invented and utilized in factories based on the development of modern computers, integrated circuits, and robotics science. The breakthrough era of industrial robots after the first batch was employed in General Motors' new factory in Ohio. Today, robots are widely used in manufacturing, extending from industrial robotic manipulators from well-known vendors like ABB, KUKA, UR, FANUC, and YASKAWA, to other kinds of robots, including AGVs and UAVs (unmanned aerial vehicles). These robotic applications include carrying, painting, packaging, polishing, and emerging human-robot collaboration. So far, robots have effectively liberated human workers from repetitive and overloaded tasks in the factory and propelled the manufacturing industry into a new era of automation. However, they have not yet surpassed the fundamental boundary of robotics, which is to manipulate the physical world under computer's control. Challenges from dynamics, uncertainties, and flexibility arise.

The advancement of robotics technology has led to the development of smart robotic manufacturing, in which robots are capable of handling more complex tasks with greater intelligence. These robots are driven by motors powered by electricity. Motors generate forces and operate the mechanical body of the robot. The moverent of the motors inside a robot allows the end effector, also known as the tool central point (TCP), to be positioned in the Cartesian coordinate system according to the rules of robotics kinematics. This enables the robot to perform various manufacturing tasks using different tools attached to the end effector. Hence, there are three levels of abstraction in robot control:

(1) Level 1: Motor level. Motor control is crucial for determining the amount of force generated by a motor. It is influenced by the power applied to a motor and affects the objects that come into direct contact with the end effectors. The speed of motor rotation

also relies on the power. In manufacturing, the force control is commonly used for tasks such as grasping, levering, or haptic human-robot collaboration, using torque sensors or sensorless estimation. In customized robotic devices, it is possible for end-users to program the motor power using motor drivers and microcontrollers (e.g., pulse-width modulation). However, the availability of motor power is usually restricted by commercial robot vendors, resulting in compromises for users. Collaborative robots like KUKA iiwa, ABB GoFa, and UR with full torque sensors offer greater possibilities.

(2) Level 2: Motion level. Motion control, also known as motion planning, plays a significant role in industrial robotic applications, from path planning of AGVs and UAVs to trajectory planning of industrial robots. Sequential postures represent these motions on a plane or in the Cartesian coordinate system on the temporal dimension. The posture of a robot is characterized by its position and gesture in the form of temporal values of joint angles. Control of the joint angles takes motor control as the foundation. Commercial industrial robots support their key functionality through programmable motion control, such as ABB RAPID and KUKA KRL robotic programming languages. The motion level is an abstraction of the motor level because it focuses more on the motion, path, or trajectories of the robot or the TCP of the robot, while the motors govern the momentary movement.

(3) Level 3: Task level. Task control involves integrating motor and motion control to create complete solutions for industrial robotic tasks such as packaging, transporting, cutting, welding, polishing, and painting. A task consists of multiple procedures, generally equating a plan of the robot's trajectory. A trajectory is a sequence of TCP postures in the Cartesian coordinate system with start and end points, which vary due to different procedures. By splicing trajectories sequentially, a robot can perform logical actions in the physical world. However, trajectories are not the only element in a task. Triggering I/O signals, alerting on the user interfaces, and writing into the database are also procedures of a task, especially when a comprehensive work cell is targeted. Overall, the task level takes motor and motion levels as an abstraction based on the control from joint angles and the torque on the trajectory.

Control of motor, motion, and task levels can be easily confused. They are different subjects that require various domains. These domains depend on multiple disciplines, including electromechanics, automatic control, production scheduling, robotic kinematics, robotic dynamics, planning algorithms, and optimization. Because of business secrets and patents, motor control of commercial industrial robots is usually concealed, with emphasis placed on the industrial users' motion and task control. However, controlling motion and tasks in a static setting cannot be considered smart or intelligent. To achieve a new generation of smart robotic manufacturing, robots need to adapt to dynamic and varied environments. Human interference renders planned trajectories unfeasible for the robot to follow. Hence, the robot needs to be flexible enough to interact with humans, such as adjusting its motion while approaching the end points. This flexibility is related not only to the motion level but also to the task level when humans change planned procedures, such as changing the order of procedures or the time required. The challenge lies in the fact that interferences present uncertainties, leading to a trickier issue: the robot requires advanced intelligence to address these stochastic properties. Machine learning is a promising information technology that can enhance intelligence through robust representation.

2. Machine learning

Machine learning is a field that combines computation and statistics, with connections to information theory, signal processing, algorithm, control theory, and optimization theory. Nowadays, machine learning has become one of the most fascinating areas of artificial intelligence. The basic idea behind machine learning is to find functions that fit data. The trained function is then used for approximation to new data. The development from a weak function to a stronger one utilizing available data is known as "learning", which is similar to the learning capabilities and processes of animals. Since this development is typically carried out by machines such as computers, the term "machine learning" was coined.

In machine learning, the fitted function takes raw data or handcrafted feature as input and generates output by passing the information through itself. Before the advent of deep learning, feature engineering involved manually extracting features from data or using predefined rules. However, this process has been replaced by more complex models, such as deep neural networks. In addrtion to neural networks, other models such as linear models, quadratic models, expressions of probability distributions (e.g., Gaussian estimation or Gaussian mixture models), and discrete tables are also common. In deep learning, alternatives to dense neural network (DNN) indude convolutional neural network (CNN), recurrent neural network (RNN), or graph neural network (GNN). The choice of models depends on the nature of problems. If a problem is easy to model or adequately dealt with by known probability distributions, these models can be selected with parameters to be learned. For tricky problems, neural networks are preferred because they can theoretically represent any non-linear functions. The accessibility of the model often depends on the dimension of the data. [1] For instance, extracting meaning from images requires an approximator with stronger presentation capability such as deep neural networks because images usually have a larger volume of data compared to standard numerical inputs. Different tasks require different neural networks; for example, CNNs are designed for processing images, while RNNs are suitable for sequential data like texts and sounds. No matter which model is selected, the learning phase involves estimating the parameters based on the given data. Elements such as the model type and inner settings (e.g., the combination of several Gaussian distributions, the number of layers in a neural network) are known as hyperparameters.

The input data can be a scalar, a vector, a matrix, or a tensor within the mathematical definitions. In reality, it can be a numerical array, a paragraph of text, a piece of audio, or a picture. The output can also have different dimensions based on the user's requirements, and sometimes it can even be the same as the input. For example, variational auto-encoder (VAE) and generative adversarial network (GAN) can produce outputs with different dimensions. Machine learning uses its function to map data from input to output, and different mappings result in different applications. For instance, mapping an image to a label is recognition; mapping a sentence from one language to another is machine translation; mapping any data from itself to itself can be used for machine generation (e.g., painter or poet) or dimension reduction.

Understanding the function parameters is essential for machine learning. Parameter tuning depends on the type of machine learning being used. There are three main types of machine learning: unsupervised learning, supervised learning, and reinforcement learning. Unsupervised learning focuses on information processing without supervision signals, such as clustering. Supervised learning provides supervision signals that can be used to calculate loss functions, and the parameters of the function are adjusted to optimize the cost function. Gradient methods are commonly used for optimization because they can be effectively calculated by computers. Reinforcement learning has a view from another point, which is based on the "trial-and-error" paradigm. Considering human learning, supervised learning is like learning from reading books, while reinforcement learning is like grasping knowledge from experience. Not all knowledge can be learned from books, and one hypothesis states that any intelligence and associated abilities can be understood by maximizng the reward (or minimizing the cost). Reinforcement learning is a crucial method for robot learning.

3. Robot learning

In the future, robots in manufacturing will not only be confined to fixed stations where they execute the same procedures thousands of times. Instead, they will be involved in changing environments where tasks cannot be fully preprogrammed. This challenge requires the robot to handle uncertainties and act accordingly.[2] Robot learning is a technology that allows robots to learn by themselves with the help of humans, using their sensors and motors to extend and enhance their initial intelligence to cope with new environments.

Robot learning is an integration of multiple machine learning technologies within the field of robotics. It stresses robot learning and acting using machine learning technology. Unlike common machine learning, robot learning emphasizes generating actions as the output while observing the environment as the input. For instance, deep learning helps the robot handle unstructured environments, while reinforcement learning provides formalisms for machine behavior. This kind of paradigm is similar to how animals behave. Humans perceive the environment by their sense organs (e. g. , eyes, nose, and hands) linked to the brain through the nervous system. For instance, the ray of light in the environment is perceived by light-sensitive cells on the retina, followed by bio-electricity signals to the cerebral cortex. The cerebral cortex also sends signals to the corresponding muscles to generate motion. This pipeline forms a close-loop style, just like automatic control. For example, when a human is going to grasp a glass of milk, they will use their eyes to locate the glass and monitor their hand while moving their arm to reach the glass, lift it, move it to their mouth, drink, and later put the glass back. Robotic devices in the closed-loop share similar behavior. The robot utilizes cameras to observe the environment and uses computers to process perception data. Algorithms then send signals to the robot controller to drive the robotic manipulator. However, no animal is born with such skills but learns show to observe, make decisions, and act during growth. Furthermore, this learning style is not fully supervised but more like learning from experience or "trial-and-error" that reinforcement learning emphasizes. Robot learning is inspired by how animals learn and adapt.

While general machine learning focuses on prediction and classification, robot learning stresses more on the output of configuration values of robotic systems, e. g. , temporal joint values or task-oriented procedures. Those signals generate motion belonging to machine behavior. This kind of technology often takes real-time data by sensors of the physical world as input, also known as observation. Policy or controller is the core mapping observation to actions indicated by robotic configurations. By doing so, one robot realizes manipulation of the physical environment and results in the state transition of the environment. Then, one new observation of the state perceived by the sensors is passed to a new iteration. Similar pipelines are often defined as the Markov decision process (MDP), a simplified mathematical model for decision-making and state transition. Since decisions are made one after another, this problem is also called sequential decision-making. Concretely, unlike machine learning supervised by a direct loss function, robot learning relies on accumulated reward (or cost) signals from the environment. Reinforcement learning plays a role of the backend mechanism when a robot learns. [3] The one who makes decisions is often called an agent who receives the observation and runs a policy. Observation varies from full observation to partial observation due to perception conditions. Policies are divided into on-policy methods and off-policy methods regarding the sampling style. The availability of environment model leads to model-based robot learning and model-free robot learning. We will review and discuss the details of those terminologies in the following sections.

Robot learning appears in various robotic skills. Learning for manipulation focuses on manipulating a vast array of objects through continuous learning. Robots are currently skilled at picking and manipulating with rigid geometry in a repetitive style but weaker at variation in movements. Purposive reasoning is another challenging issue, especially in the interaction with humans, which hammers at reasoning human activities for enhanced human-robot collaboration. Robot learning with data-driven approximation (e.g., neural networks) shows excellent capability in robot grasping, agile motor control, motion planning, identifying by feeling, and so forth. Learning from imitation is also a classic but popular research topic, e.g., learning from human teachers, performance, or other tasks that are beyond imitation. Those robotic skills also show significant potential in manufacturing to shift from mass production to mass customization. In smart robotic manufacturing, robot learning has already shown its initial strength on motor, motion, and task levels.

22.2 Words and Phrases

smart robotic manufacturing	智能机器人制造
industrial robots	工业机器人
robotic manipulators	机器人操纵器
AGVs (automated guided vehicles)	自动导引车
UAVs (unmanned aerial vehicles)	无人机
human-robot collaboration	人机协作
end effector	末端执行器
TCP (tool central point)	工具中心点
motor level	电机控制水平
motor control	电机控制
torque sensor	转矩传感器
senseless estimation	无传感器估计
motion level	运动控制水平
motion planning	运动规划
trajectory planning	轨迹规划
task level	任务控制水平
task control	任务控制
robot dynamics	机器人动力学
robot kinematics	机器人运动学
electromechanics	电机机械
automatic control	自动控制
production scheduling	生产排程
planning algorithms	规划算法
machine learning	机器学习
supervised learning	监督学习
unsupervised learning	无监督学习

reinforcement learning　　　　　　强化学习
robot learning　　　　　　　　　　机器人学习
Markov decision process（MDP）　　马尔可夫决策过程

22.3　Complex Sentence Analysis

［1］ For instance, extracting meaning from images requires an approximator with stronger presentation capability such as deep neural networks because images usually have a larger volume of data compared to standard numerical inputs.

① 主语是 extracting meaning from images，这是一个动名词短语作主语，表示"从图像中提取意义"这一动作或过程。

② 宾语补足语（修饰宾语的部分）是 with stronger presentation capability such as deep neural network，这是一个介词短语修饰宾语 approximator，用来描述这个近似器应具备的特性。

［2］ Robot learning is a technology that allows robots to learn by themselves with the help of humans, using their sensors and motors to extend and enhance their initial intelligence to cope with new environments.

主句为 Robot learning is a technology that allows robots to learn by themselves with the help of humans。

从句为目的状语从句，描述了主句中的技术的目的，修饰了主句中的 Robot learning。

［3］ The one who makes decisions is often called an agent who receives the observation and runs a policy.

① 主语是 The one who makes decisions，这是一个由定语从句修饰的名词短语，表示"作出决策的人"。

② 定语从句是 who receives the observation and runs a policy，关系代词 who 指代 agent；谓语是 receives 和 runs；宾语是 the observation 和 a policy。

22.4　Exercise

Translate the following paragraph

Robotic equipment has been playing a central role since the proposal of smart manufacturing, showing potential to enhance factory automation. Beginning with the first integration of industrial robots into production lines, robots like industrial manipulators, collaborative robots, automated guided vehicles（AGVs）, and even unmanned aerial vehicles（UAVs）have been employed to beef up manufacturing systems from one generation to another. With the development of artificial intelligence（AI）like deep learning and computing devices such as GPU (graphic processing unit) and NPU (neural processing unit), robots are expected to have enhanced intelligence. This trend focuses on releasing the robots from the clampdown of old-fashioned hard-programmed scenarios and investigating opportunities where the robots can qualify for more flexible and human-centered tasks. Aimed at more complex and flexible environments, robots are supposed to master advanced skills by self-learning to achieve success, just like machine learning.

Lesson 23　Product Test and Quality Control

23.1　Text

1. Product Test

Product test can be thought of as the culmination of all process control work. It can also be considered as quality check of the inspection process itself. If the quality plan is adequate and carried out properly, then the product's performance should be verified, making a total test redundant. For this reason, a test of the completed product is often nothing more than a contractual requirement that must be performed before customers accept the product. However, product test is more than just proving. Testing the whole is equal to the sum of the parts. [1] It allows for gathering data that supports the design theory of the product, for interpretations to be made for further improvements in design so that future products will be better than present ones, and for evaluation of design evolution toward better performance and costs. In addition, it is a means of verifying design, since not all design parameters can be fully calculated or predicted.

Product test engineers work closely with design engineers to provide useful data for testing. They must also work in close harmony with engineers in all other phases of manufacturing. Frequently, product test will reveal deficiencies in design that requires major revisions in manufacturing processes. This is particularly true if the company produces many prototypes and has short production runs. Therefore, manufacturing engineers are as interested in product test results as are design engineers.

For complex products, product test becomes a very important part of the total process control function. It gives the company a high degree of confidence that the product will perform as the customer expects it to, and this is a valuable marketing tool as it helps to establish a proper reputation with customers.

2. Geometric Errors

Geometric errors are defined here as errors in form of individual machine components (e.g., straightness of motion of a linear bearing). Geometric errors are concerned with the quasi-static accuracy of surfaces, which impact the moving relationships of surfaces. Geometric errors can be smooth and continuous (systematic) or can exhibit hysteresis (e.g., backlash) or random behavior. [2] Many factors affect geometric errors, including surface straightness (Figure 23.1), surface roughness, bearing preload, kinematics versus elastic design principles, and structural design philosophies.

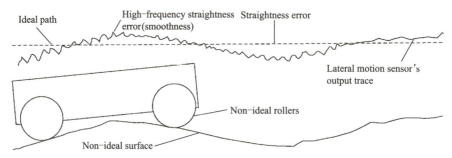

Figure 23.1　Geometri Errors Caused by Surface Straightness

3. Quality Planning

Quality planning is the planning and strategy activity of process control, and it is sometimes referred to as process planning or inspection planning. The engineers involved develop plans for checking the adequacy of performance of shop operations to ensure that the final product performs as designed. Using plans and methods produced by measurement of productivity and work measurement (MP&WM) as a guide, quality planning engineers determine where inspections and nondestructive tests will be specified during the manufacturing process. They also specify the type of inspection or test to be conducted and, based on design engineering requirements, determine what will constitute acceptance or rejection.

Normal manufacturing activities produce a certain percentage of deviations from drawings. Some are important, some are of little consequence. It is the quality planning engineers' responsibility to evaluate these deviations and determine what the proper corrective action will be. They then through MP&WM ensure that the corrective actions are factored into manufacturing planning for rework. As the arbiter of quality via the deviation from drawing procedure, quality planning has the database to evaluate performance of the various shop and support functions. A score-keeping function is possible and desirable; in this way, quality planning can report to management whether quality levels are improving or declining.

Quality state can be reported by statistically evaluating the numbers of deviations and their seriousness. This leads naturally to an evaluation of the cost of doing the repair work caused by the deviation. Repair work, which constitutes manufacturing losses, is an important measurement of organizational quality levels. Manufacturing losses can be used to measure the adequacy of attention to detail of the operators and their foremen. High losses indicate a poorly managed operation. Quality planning engineers are responsible for setting the manufacturing losses, budgets, and measurements policy.

4. Quality Control

Quality control has traditionally been the liaison between manufacturing and design.

Product Test and Quality Control Lesson 23

This function interprets design specifications for manufacturing and develops the quality plan to be integrated into manufacturing engineering's methods and planning instructions to operations. Quality control is also responsible for recommending to management what level of manufacturing losses (cost of mistakes in producing the product) can be tolerated. This is based on the complexity of the product design, specifically the degree of precision necessary in tolerances. Quality control traditionally monitors manufacturing losses by setting a negative budget that is not to be exceeded and establishes routines for measurement and corrective action.

Within the past decade or two, quality control has become increasingly involved with marketing and customers in establishing documentation systems to ensure guaranteed levels of product quality. This new role has led to the new title "quality assurance", differentiating it from traditional in-house quality control.

Quality assurance strives, through documentation of performance and characteristics at each stage of manufacture, to ensure that the product will perform at an intended level. Whereas quality control is involved directly with manufacturing operations, quality assurance is involved with the customer support responsibilities generally found within the marketing function. Many industrial organizations have chosen to establish an independent quality assurance sub-function within the manufacturing function and have placed the technical responsibilities of quality control, namely process control, within the manufacturing engineering organization.

23.2 Words and Phrases

culmination	*n.* 顶点
contractual	*adj.* 契约的
deficiency	*n.* 缺乏，不足
quasi-static	准静态的
hysteresis	*n.* 滞后作用，[物] 磁滞现象
liaison	*n.* 联络，(语音) 连音

23.3 Complex Sentence Analysis

[1] It allows for gathering data that supports the design theory of the product, for interpretations to be made for further improvements in design so that future products will be better than present ones, and for evaluation of design evolution toward better performance costs.

两个 for 引导的介词短语在句中充当状语。

[2] Many factors affect geometric errors, including surface straightness (Figure 23. 1), surface roughness (Figure 23. 1), bearing preload, kinematics versus elastic design principles, and structural design philosophies.

① surface straightness：表面直线度。
② surface roughness：表面粗糙度。
③ versus elastic design principles 为介词短语，修饰 kinematics。

23. 4 Exercise

Translate the following paragraphs

Machine parts are manufactured, so they are interchangeable. In other words, each part of a machines or mechanisms is made to a certain size and shape, so it will fit into any other machines or mechanisms of the same type. To make the parts interchangeable, each individual part must be made to a size that will fit the mating part correctly. It is not only impossible but also impractical to make many parts to an exact size. This is because machines are not perfect, and the tools become worn. A slight variation from the exact size is always allowed. The amount of this variation depends on the kind of part being manufactured. For example, a part might be made 6in (about 0.15m), along with a variation allowed of 0.003in (about 0.000076m) above or below this size. These are known as the limits. The difference between upper and lower limits is called the tolerance. A tolerance is the total permissible variation in the size of a part. The basic size is limited by the application of allowances and tolerances.

Sometimes the limit is allowed in only one direction. This is known as unilateral tolerance. Unilateral tolerance is a system of dimensioning where the tolerance (that is, variation) is shown in only one direction from the nominal size. Unilateral tolerance allows the changing of tolerance on a hole or shaft without seriously affecting the fit. When the tolerance is in both directions from the basic size, it is known as a bilateral tolerance (plus and minus). Bilateral tolerance is a system of dimensioning where the tolerance (that is, variation) is split and shown on either side of the nominal size. Limit dimensioning is a system of dimensioning where only the maximum and minimum dimensions are shown. Thus, the tolerance is the difference between these two dimensions.

Lesson 24 Mechatronics

24.1 Text

Mechatronics was originally coined in the 1970s from the integration of two engineering disciplines—mechanics and electronics. [1] More recently, with spectacular advancements in the areas of control and communications, the word "mechatronics" has been adapted as the synergetic integration of three disciplines: mechanics, control, and electronics, aiming at the study of mainly manufacturing machines controlled by electronics. It is also being viewed as the fusion of mechanical engineering with electronics and intelligent computer control in the design and manufacture of industrial products and processes. Lately, the notion of "intelligence" or "smartness" is associated with these industrial machines, and they are being called "intelligent" or "smart" machines. Mechatronics is also defined as the synergetic integration of mechanical engineering with electronics and intelligent computer control in the design and manufacture of products and processes.

[2] In engineering terms, what can be made to emerge is a new and previously unattainable set of performance characteristics. Thus, mechatronics is truly an interdisciplinary subject drawn from mechanical, electrical, electronics, computer, and manufacturing engineering. The subject of mechatronics is gaining acceptance as an essential course in engineering education, and the field of mechatronics is a basis for new industrial development. The technology areas of mechatronics involve system modelling, simulation, sensors and measurement systems, drive and actuation systems, analysis of the behavior of systems, control systems, and microprocessor systems.

1. Mechatronics paradigm

[3] The mechatronics paradigm deals with benchmarking and emerging problems in engineering, science, and technologies which have not been addressed and solved. Mechatronics is an integrated comprehensive study of intelligent and high-performance electromechanical systems (mechanisms and processes), intelligent and motion control through the use of advanced microprocessors and DSPs (digital signal processors), power electronics and ICs (integrated circuits), design and optimization, modelling and simulation, analysis and virtual prototyping, etc. Integrated multidisciplinary features are quickly approaching, and mechatronics, which integrates electrical, mechanical, and

computer engineering areas (Figure 24.1), takes place.

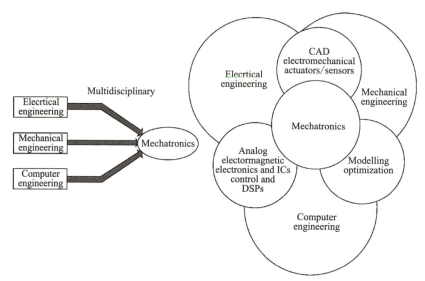

Figure 24.1 Mechatronics Integrates Electrical, Mechanical, and Computer Engineering

One of the most challenging problems in mechatronics systems design is the development of system architecture, e.g., selection of hardware (actuators, sensors, devices, power electronics, ICs, microcontrollers, and DSPs) and software (environment and computation algorithms to perform sensing and control, information flow and data acquisition, simulation, visualization, and virtual prototyping).[4] Attempts to design state-of-the-art man-made mechatronics systems and to guarantee the integrated design can be pursued through analysis of complex patterns and paradigms of evolutionary developed biological systems. Recent trends in engineering have increased the emphasis on integrated analysis, design, and control of advanced mechanical systems. The scope of mechatronics system has continued to expand and include actuators, sensors, power electronics, ICs, microprocessors, DSPs, as well as I/O devices.

The mechatronics paradigm was introduced with the ultimate goals to:

(1) Guarantee an eventual consensus and ensure descriptive multidisciplinary features.

(2) Extend and augment the results of classical mechanics, electromechanical systems, power electronics, ICs, and control theory to advanced hardware and software.

(3) Acquire and expand the engineering core integrating interdisciplinary areas.

(4) Link and place the integrated perspectives of electromechanical systems, power electronics, ICs, DSPs, control, signal processing. MEMS, and NEMS in the engineering curriculum in favor of the common structure needed.

The study of high-performance electromechanical systems should be considered as the unified cornerstone of the engineering curriculum through mechatronics. The unified analysis of actuators and sensors (e.g., electromechanical motion devices), power electronics and ICs, microprocessors and DSPs, advanced hardware and software, have barely been introduced in the engineering curriculum. Mechatronics, as a breakthrough concept in design and analysis of conventional, micro-scale and nano scale electromechanical systems, was introduced to address, integrate, and solve a great variety of emerging problems.

2. Classification of Mechatronics

Mechatronics system can be classified as conventional mechatronics system, microeletromechanical system (MEMS), and nano-electromechanical system (NEMS). The operational principles and basic fundamentals of conventional mechatronics system and MEMS are the same, while NEMS is studied using different concepts and theories. In particular, the designer applies the classical mechanics and electromagnetics to study conventional mechatronics systems and MEMS. Quantum theory and nano-electromechanics are applied in NEMS. The fundamental theories used to study the effects, processes, and phenomena in conventional, micro-scale and nano-scale mechatronics systems are illustrated in Figure 24.2.

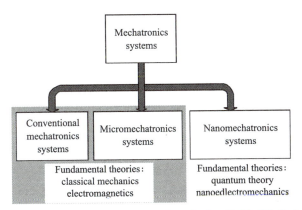

Figure 24.2 Classification and Fundamental Theories Applied in Mechatronics Systems

24.2 Words and Phrases

mechatronics	n. 机电一体化
coin	v. 造词
integration	n. 集成
synergetic	adj. 协同的，合作的
fusion	n. 融合
notion	n. 概念，想法
unattainable	adj. 做不到的
interdisciplinary	adj. 学科间的
paradigm	n. 范例
benchmark	n. 基准，标准

entreaty	*n.* 恳求，请求
evolutionary	*adj.* 发展的，演化的
DSP（digital signal processing）	数字信号处理
IC（integrated circuit）	集成电路
consensus	*n.* 一致
augment	*v.* 增加，扩大
unify	*v.* 成为一体，统一
cornerstone	*n.* 基石，基础

24.3　Complex Sentence Analysis

［1］ More recently, with spectacular advancements in the areas of control and communications, the word "mechatronics" has been adapted as the synergetic integration of three disciplines: mechanics, control, and electronics, aiming at the study of mainly manufacturing machines controlled by electronics.

① with：随着。
② aiming at：目的为。

［2］ In engineering terms, what can be made to emerge is a new and previously unattainable set of performance characteristics.

① In engineering terms：从工程上来说。
② what can be made to emerge 是主语从句。

［3］ The mechatronics paradigm deals with benchmarking and emerging problems in engineering, science, and technologies which have not been addressed and solved.

① deals with：处理。
② which have not been addressed and solved 是定语从句，修饰 problems。

［4］ Attempts to design state-of-the-art man-made mechatronics systems and to guarantee the integrated design can be pursued through analysis of complex patterns and paradigms of evolutionary developed biological systems.

① attempts to…：在……方面的尝试，在……方面的努力。
② state-of-the-art man-made：具有艺术特性的人造的。

24.4　Exercise

Translate the following paragraphs

A mechatronics product, such as modern SLR (single lens reflex) cameras, video recorders, music synthesizers, and automobiles with engine management systems, cannot be designed by a single person simply because it is too complex. Nor can it be designed by a large number of persons with different specializations unless these persons operate in a team manner.

It used to be the case where a product containing some mechanical functionalities and electronic sensing and control would be designed sequentially. Firstly, the basic mechanical structure would be designed and made by the mechanical engineers. Then, it would become the job of the electronics engineers to fit it with the appropriate transducers and actuators. Finally, the control engineers would be given the job of finding effective controller structures and algorithms to drive it. There are a number of the disastrous consequences of this design philosophy when the product is of a high degree of complexity.

Instead, the mechatronics design methodology demands teamwork right from the start, with the team including not only the technical design experts from the various contributing disciplines but also from marketing, financial, and other departments.

Lesson 25　Introduction to MEMS

25.1　Text

1. Development of MEMS

MEMS was only an assumption in the early 1980s. [1] Only when the application of manufacturing technology (semi-conductor microprocessing technology) in the micro-scale product field created the condition to develop MEMS, and much relative technology such as design, material, measuring, control, sensing, information processing, computer, energy, and system integration achieved a certain high-level achievement, i.e., in the late 1980s, researchers at the University of California, Berkeley, and MIT successfully researched and manufactured electrostatic micro-motor with a diameter of 100μm, did MEMS begin to be popularly studied and highly regarded worldwide, becoming one of the newly emerging high technologies.

Since it was predicted that micrometer/nanometer technology would lead to an industry revolution, many developed countries and areas viewed it as a crucial technology in both economic flourishing and national defense. They prioritized it as a project and spent huge sums of money researching it, which made it rapidly develop and achieve certain milestones. For example, Stanford University researched and manufactured gemel connecting rod level mechanism with a diameter of 20μm and a length of 150μm, a sliding block mechanism with a size of 210μm × 100μm, a micro electrostatic motor with a diameter of 200μm, and a pump with a flux of 20mL/min. Tokyo University researched and manufactured a micro slope climbing mechanism with a size of 1cm. [2] Nagoya University researched and manufactured a crawling wireless robot used to inspect micro pipes, with its movement controlled by a magnetic field generated by a circuit loop outside the pipe, etc. At the same time, MEMS is also highly-regarded in our country. Now, the miniaturization technology on mechanical parts, integrated sensor, optics parts, and actuator has been studied. Some of them have achieved certain milestones and have turned to application studies. Scientists from Shanghai Jiao Tong University have developed a multi-stage delivery strategy for magnetically controlled micro-nano robots based on the humanorgan-on-a-chip (OOC) model they built. With the help of micro-nano 3D printing technology, they built a layered vascularized OOC model that can reproduce the structure and function of different organs in the human body. The team can inject magnetically guided micro-nano robots into these OOCs and navigate them within the blood vessels using external magnetic fields. Applications of these in vitro models include promoting drug screening, reducing the cost of drug research, and improving the efficiency of drug development. They can also be used in medical treatment, such as establishing in vitro OOCs for tumor patients as a substitute for patients to try different drugs and guide clinical medication. To solve the needs of micro-parts processing, Sukode has innovatively developed the ultra-high precision equipment KASITE-SKD series micro nano processing center with a processing allowance range from 20nm to 100μm.

2. Basic Characteristics

The miniaturizing of MEMS causes the scaling effect problem, i.e., the physical phenomena do not proportionally miniaturize the size. When the size is sufficiently small, the simulation and similarity theory of macro-mechanical is no longer adaptive. Therefore, MEMS has the following basic characteristics:

(1) Dominant force.

The dominant force of MEMS is a surface force. We know the bulk force (e.g., gravity, electromagnetic force) is direcly proportional to high power feature size, while the surface force (friction, surface, and electrostatic force) is directly proportional to a relatively low power of feature size. MEMS has small bulk and is lightweight. The ratio of surface force to bulk force relatively increases. Compared with bulk force, the surface force becomes the dominant force. After miniaturizing, compared to gravity, electrostatic force becomes the dominant force (the scaling effect of mechanical miniaturization). Therefore, MEMS is often actuated by electrostatic force. Compared to gravity, the effect of friction in MEMS is also larger than that in a normal machine.

(2) Non-simulative miniatwrization.

MEMS is not the simulative miniaturization of a traditional machine. The complexity of every part of traditional mechanical systems is not equal. To geometrically miniaturize them, the miniaturization of complicated parts is very difficult, especially in highly intelligent automotive mechanics. To design MEMS, there is no need to pursue complicated mechanical structures but rather focus on multiple single mechanical parts (include parts with sensors and artificial intelligence) which can complete complex work.

(3) Energy supply.

For MEMS which can move and spin, electrical cables are obstacles to the movement. Energy supplying usually does not require cables. Now, MEMS often supplies energy by electrostatic force. Besides, MEMS is often directly actuated by vibration (piezoelectric, electromagnetic, and SMA actuating).

Therefore, the development of MEMS needs new theories and methods. With the change of main resistance, new structure principles and control methods are needed. With the change of the dominant elements in kinematic and kinetic equations, new actuating methods are needed. With the miniaturization of the structure of MEMS parts, new manufacturing methods are needed, etc.

3. Applications of MEMS

The rise and development of MEMS synchronously forecast its wide application prospects. The main fields can be concluded as follows:

(1) Machine field: micro gearing, micro connecting rod level mechanism, micro sliding block mechanism, etc., which are composed of micro systems.

(2) Instruments: pressure sensors, acceleration sensors, etc.

(3) Hydro-control: micropumps, intelligent pumps, etc.

(4) Micro optics: optical cables, optical scanners, interferometers, etc.

(5) GSI: vacuum manipulators, micro position systems, gas precision control systems, etc.

(6) Information machines: magnetic heads, printer heads, scanners, etc.

(7) Robots in the next century: microrobots, multi-degree-of-freedom manipulators, etc., which can be used in minimal environments such as micro pipe inspection and repair.

25.2 Words and Phrases

micro-processing technology	微加工技术
electrostatic	*adj.* 静电的，静电学的
micrometer/nanometer technology	微/纳米技术
flourishing	*adj.* 繁茂的，繁荣的，欣欣向荣的
simulative	*adj.* 模拟的，假装的
miniaturize	*vt.* 使小型化，使微型化
synchronous	*adj.* 同时的，[物] 同步的
micro gearing	微齿轮
micro connecting rod level mechanism	微连杆机构
micro sliding block mechanism	微滑块机构

25.3 Complex Sentence Analysis

[1] Only when the application of manufacturing technology…, did MEMS begin to be popularly studied and highly regarded worldwide, becoming one of the newly emerging high technologies.

当 only ＋从句或短语提前时，主句需要倒装。

试比较：

a. Only when we broaden our views can we realize the importance of knowledge.

b. Only in this way can we deal with the matter effectively.

[2] Nagoya University researched and manufactured a crawling wireless robot used to

inspect micro pipes, with its movement controlled by a magnetic field generated by a circuit loop outside the pipe, etc.

① used to inspect micro pipes 为过去分词短语作定语，修饰前面的 robot。
② with its movement controlled by a magnetic field generated by a circuit loop outside the pipe, etc. 为介词短语作状语，解释了机器人的运动是如何实现的。

25.4　Exercise

Translate the following paragraphs

MIT is developing a MEMS-based gas turbine generator. Based on high-speed rotating machinery, this 1cm diameter by 3mm thick Si heat engine is designed to produce 10—20W of electric power while consuming 10g/h of H_2. Later versions may produce up to 100W using hydrocarbon fuels. The combustor is now operating, and an 80W micro turbine has been fabricated and is being tested. This engine can be considered the first of a new class of MEMS devices, power MEMS, which are heat engines operating at power densities similar to those of the best large-scale devices made today.

The design of the micro-gas turbine generator presents a considerable challenge to all the disciplines involved. However, progress to date has been quite encouraging. The ability to manufacture MEMS-based, high-speed rotating machinery opens up a host of possibilities, including various thermodynamic machines. MIT is also working on a motor-driven micro compressor and a micro-high pressure liquid rocket motor employing turbo pumps. The concept of MEMS-based, high power density heat engines appears extremely attractive and physically realizable.

Lesson 26 Industrial Robots

26.1 Text

[1] A robot is an automatically controlled, reprogrammable, multipurpose, manipulating machine with several reprogrammable axes which may be either fixed in place or mobile for use in industrial automation applications. The key words are reprogrammable and multipurpose because most single-purpose machines do not meet these two requirements. The term "reprogrammable" implies two things: The robot operates according to a written program, and this program can be rewritten to accommodate a variety of manufacturing tasks. The term "multipurpose" means that the robot can perform many different functions, depending on the program and tooling currently in use.

Over the past two decades, the robot has been introduced in the manufacturing industry to perform many monotonous and often unsafe operations. Because robots can perform certain basic tasks more quickly and accurately than humans, they are being increasingly used in various manufacturing industries.

1. Structures of Industrial Robots

The structure of industrial robot consists of four major components: the manipulator, the end effector, the power supply, and the control system, as shown in Figure 26.1.

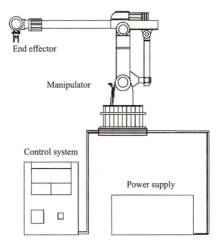

Figure 26.1 The Structure of Industrial Robot

The manipulator is a mechanical unit that provides motions similar to those of a human arm. It often has a shoulder joint, an elbow, and a wrist. It can rotate or slide, stretch out and withdraw in every possible direction with certain flexibility.

The basic mechanical configurations of the robot manipulator are Categorized as Cartesian, cylindrical, spherical, and articulated. A robot with Cartesian geometry can move its gripper to any position within the cube or rectangle defined as its working volume. Cylindrical coordinate robots can move the gripper within a volume described by a cylinder. The cylindrical coordinate robot is positioned in the work area by two linear movements in the X and Y directions and one angular rotation about the Z axis. Spherical arm geometry robots position the wrist through two rotations and one linear actuation. Articulated industrial robots have an irregular work envelope, with two main variants: vertically articulated and horizontally articulated.

The end effector attaches to the end of the robot wrist, also called end-of-arm tooling. [2] It is a device intended for performing the designed operations as a human hand can. End effectors are generally custom-made to meet special handling requirements. Mechanical grippers are the most commonly used and are equipped with two or more fingers. The selection of an appropriate end effector for a specific application depends on factors such as payload, environment, reliability, and cost.

The power supply is the actuator for moving the robot arm, controlling the joints, and operating the end effector. The basic types of power sources include electrical, pneumatic, and hydraulic. Each source of energy and each type of motor has its own characteristics, advantages, and limitations. An AC-powered or DC-powered motor may be used depending on the system design and applications. These motors convert electrical energy into mechanical energy to power the robot. Most new robots use an electrical power supply. Pneumatic actuators have been used for high speed, non-servo robots and are often used for powering tooling such as grippers. Hydraulic actuators have been used for heavier lift systems, typically where the accuracy is not required.

The control system is the communications and information-processing system that gives commands for the movements of the robot. It is the brain of the robot; it sends signals to the power source to move the robot arm to a specific position and to actuate the end effector. It is also the nerves of the robot; it is reprogrammable to send out sequences of instructions for all movements and actions to be taken by the robot.

An open-loop control system is the simplest form of the control system, which controls the robot only by following predetermined step-by-step instructions. This system does not have a self-correcting capability. A close-loop control system uses feedback sensors to produce signals that reflect the current states of the controlled objects. By comparing those feedback signals with the values set by the programmer, the close-loop control system can direct the robot to move to the precise position and assume the desired attitude, and the end effector can perform with very high accuracy as the close-loop control system can minimize the discrepancy between the controlled object and the predetermined references.

【拓展视频】

【拓展视频】

2. Classification of Industrial Robots

Industrial robots vary widely in size, shape, number of axes, degrees of freedom, and design configuration. Each factor influences the dimensions of the robot's working envelope or the volume of space within which it can move and perform its designated task. A broader classification of industrial robots can been described as below:

(1) Fixed-sequence and variable-sequence industrial robots.

The fixed-sequence industrial robot (also called a pick-and-place industrial robot) is programmed for a specific sequence of operations. Its movements are from point to point, and the cycle is repeated continuously. The variable-sequence industrial robot can be programmed for a specific sequence of operations but can be reprogrammed to perform another sequence of operation.

(2) Playback industrial Robot.

An operator leads or walks the playback industrial robot and its end effector through the desired path. The robot memorizes and records the path and sequence of motions and can repeat them continually without any further action or guidance by the operator.

(3) Numerically Controlled industrial Robot.

The numerically controlled industrial robot is programmed and operated much like an NC machine tool. The robot is servo-controlled by digital data, and its sequence of movements can be changed with relative ease.

(4) Intelligent industrial Robot.

【拓展视频】

[3] The intelligent industrial robot is capable of performing some of the functions and tasks carried out by human beings. It is equipped with a variety of sensors with visual and tactile capabilities.

3. Applications of Industrial Robots

【拓展视频】

Industrial robots are a very special type of production tool, and their applications are quite broad. These applications can be grouped into three categories: material processing, material handling, and assembly.

In material processing, industrial robots use tools to process raw materials. For example, the robot tools could include a drill, allowing the robot to perform drilling operations on raw materials.

【拓展视频】

Material handling consists of loading, unloading, and transferring workpieces in manufacturing facilities. These operations can be performed reliably and repeatedly with industrial robots, thereby improving quality and reducing scrap losses.

Assembly is another large application area for industrial robots. An automatic assembly system can incorporate automatic testing, robot automation, and mechanical handling to reduce labor costs, increase output, and eliminate manual handling concerns. Figure 26.2 shows SCARA robot for automatic assembly.

Figure 26.2　SCARA Robot for Automatic Assembly

26.2　Words and Phrases

reprogrammable	*adj.* 可重复编程的
manipulate	*v.* （熟练地）操作，使用（机器等），操纵
accommodate	*v.* 使适应
monotonous	*adj.* 单调的，无变化的
elbow	*n.* 肘
wrist	*n.* 手腕，腕关节
stretch out	*v.* 伸出
Cartesian	*adj.* 笛卡儿的
cylindrical	*adj.* 圆柱的，柱面的
spherical	*adj.* 球状的
articulated	*adj.* 铰接的，有关节的
gripper	*n.* 夹持器，手爪
actuation	*n.* 活动，激励，动作
envelope	*n.* [数] 包迹，包络线
variate	*n.* [数] 变量
custom-made	*adj.* 定做的，定制的
payload	*n.* 有效载荷
pneumatic	*adj.* 气动的
discrepancy	*n.* 相差，差异，矛盾
designate	*v.* 指明，指出

servo control   *n.* 伺服控制，随动控制
tactile   *adj.* 触觉的，有触觉的

26.3　Complex Sentence Analysis

[1]　A robot is an automatically controlled, reprogrammable, multipurpose, manipulating machine with several reprogrammable axes which may be either fixed in place or mobile for use in industrial automation applications.

which 引导定语从句，在从句中作主语，修饰 reprogrammable axes。

[2]　It is a device intended for performing the designed operations as a human hand can.

① intended for：用来，目的在于；是过去分词短语作定语，修饰 device。
② as a human hand can 是方式状语从句。

[3]　The intelligent robot is capable of performing some of the functions and tasks carried out by human beings.

① be capable of doing something：能够做某事。
② carried out：完成；是过去分词作定语，修饰 functions and tasks。

26.4　Exercise

Translate the following paragraphs

Although robots have always been a fantastic subject, producing iconic scenes in movies like "Star Wars" and even more fantastic kinds described in science fiction books, we find that modern robots often surpass these in reality, capacity, and development.

Humanized robots are possible if a demand for them. The only reason we do not have more advanced robots is not because of the inability to develop them, but because a specific need has not been identified. As that need becomes obvious, they will appear in larger numbers.

The future development of robotics depends largely on young and young-at-heart scientists who are less conservative, have active and imaginative minds, and have not learned to think in terms of "not practical" or "not possible". What robots can do around the home, office, factory, and other places remains to be "seen" in their minds. These innovators will create more wonderful inventions and adaptations than we have ever dreamed of. Thus, the future of robotics belongs to the young and the young-at-heart.

Lesson 27　An Army of Small Robots

27.1　Text

A group of terrorists has stormed into an office building and taken an unknown number of people hostage. They have blocked the entrances and covered the windows. No one outside can see how many they are, what weapons they carry or where they are holding their hostages. But suddenly a SWAT team bursts into the room and captures the assailants before they can even grab their weapons. How did the commandos get the information they needed to move so confidently and decisively?

【拓展视频】

The answer is a team of small, coordinated robots. They infiltrated the building through the ventilation system and methodically moved throughout the ducts. Some were equipped with microphones to monitor conversations, others with small video cameras, still others with sensors that sniffed the air for chemical or biological agents. Working together, they radioed this real-time information back to the authorities.

The challenge was to develop tiny reconnaissance robots that soldiers could carry on their backs and scatter on the floor like popcorn.[1] On the home front, firefighters and search and rescue workers could toss these robots through windows and let them scoot around to look for trapped victims or sniff out toxic materials.

1. Ant Army

In principle, diminutive robots have numerous advantages over their bulkier cousins. They can crawl through pipes, inspect collapsed buildings, and hide in inconspicuous niches. A well-organized group of them can exchange sensor information to map objects that cannot be easily comprehended from a single vantage point. They can come to the aid of one another to scale obstacles or recover from a fall. Depending on the situation, the team leader can send in a larger or smaller number of robots. If one robot fails, the entire mission is not lost because the rest can carry on.

【拓展视频】

【拓展视频】

However, diminutive robots require a new design philosophy. They do not have the luxury of abundant power and space like their larger cousins, and they cannot house all the components necessary to execute a given mission. Even carrying something as compact as a video camera can nearly overwhelm a small robot.[2] Consequently, their sensors, processing power, and physical strength must be distributed among several robots, which must then work in unison. Such robots are like ants in a colony: weak and vulnerable on their own but highly effective when they join forces.

2. Localization

One vital task that requires collaboration is localization: figuring out the team's position. Larger robots have the luxury of several techniques to ascertain their positions, such as global positioning system (GPS) receivers, fixed beacons, and visual landmark recognitions. Moreover, they have the processing power to match current sensor information to existing maps.

None of these techniques works reliably for miniature robots. They have a limited sensor range; the millibot sonar can measure distances up to about 2m. They are too small to carry GPSs. Dead reckoning—tracking position by measuring wheel speed—is hindered by their low weight.[3] Something as seemingly inconsequential as the direction of the weave of a rug can dramatically influence their motion, making odometry readings inaccurate, just as a car's odometer would fail to give accurate distances if driven on an ice-covered lake.

Therefore, we have had to develop a new technique. What we have created is a miniaturized version of GPS. By alternating their transmitting and listening roles, the robots figure out the distances among them. Each measurement takes about 30ms to complete.[4] The team leader—either the home base or a larger robot that deployed the millibots—collects all the information and calculates robot positions using trilateration. The advantage of this localization method is that the millibots do not need fixed reference points to navigate. They can enter an unfamiliar space and survey it on their own. During mapping, a few selected millibots serve as beacons. These robots remain stationary while the others move around, mapping and avoiding objects while measuring their position relative to the beacons. When the team has fully explored the area around the beacons, the robots switch roles. The exploring robots position themselves as beacons, and the previous set begins to explore. This technique is similar to the children's game of leapfrog, and it can be executed without human's intervention.

3. Tank Chain

Obstacles present small robots with another reason to collaborate. Due to their size, small robots are susceptible to the random clutter that pervades our lives. They must deal with rocks, dirt, and loose paper. The standard millibot has a clearance of about 15mm, so a pencil or twig can stop it in its tracks. To overcome these limitations, we have developed a newer version of millibots that can couple together like train cars. Each of these new millibots, about 11cm long and 6cm wide, looks like a miniature tank. Typically, they roam around independently and are versatile enough to get over small obstacles. However, when they need to cross a ditch or scale a flight of stairs, they can link up to form a chain.

What gives the chain its versatility is the coupling joint between millibots. Unlike a train coupling or a trailer hitch on a car, the millibot coupling joint contains a powerful motor that can rotate the joint up or down with enough torque to lift several millibots. To climb a stair, the chain first pushes up against the base of the stair. One of the millibots near the center of the chain then cantilevers up the front part of the chain. Those millibots that reach the top can then pull up the lower ones. Right now, this process has to be remotely controlled by humans, but eventually, the chain should be able to scale stairs automatically.

Already, researchers' attention has begun to shift from hardware development toward the design of better control systems. The emphasis will move from the control of a few individuals to the management of hundreds or thousands—a fundamentally different challenge that will require expertise from related fields such as economics, military logistics, and even political science.

One of the ways we envision large-scale control is through hierarchy. Much like the military, robots will be divided into smaller teams controlled by a local leader. This leader will be responsible to a higher authority. Already millibots are being directed by larger, tank-like robots whose processors can handle the complex calculations of mapping and localization. These larger robots can tow a string of millibots behind them like ducklings and, when necessary, deploy them in an area of interest. They themselves report to larger all-terrain-vehicle robots in the group, which have multiple computers, video cameras, and GPSs. The idea is that the larger robots will deploy the smaller ones in areas that they cannot access themselves and then remain nearby to provide support and direction.

To be sure, small robots have a long way to go. Outside of a few laboratories, no small-robot teams are roaming the halls of buildings searching for danger. Although the potential of these robots remains vast, their current capabilities place them just above novelty. As the technology filters down from military applications and others, we expect the competence of small robot to improve significantly. Working as teams, they have a full repertoire of skills; their modular design allows them to be customized to particular missions; and, not least, they are fun to work with.

27.2 Words and Phrases

special weapons and tactics (SWAT)	n. 特警队
ventilation	n. 通风
microphone	n. 麦克风
diminutive	adj. 极小的，微型的
processing power	处理能力

global positioning system（GPS）	n. 全球定位系统
midget	adj. 极小的
millibot	n. 毫米机器人
ultrasonic	n. 超声的，超声波
odometry	n. 测程法
odometer	n. 里程计
trilateration	n. 三边测量
intervention	n. 干预
cantilever	n. 悬臂梁
military logistics	军事后勤学
political science	政治学
hierarchy	n. 分级；等级

27.3　Complex Sentence Analysis

[1] On the home front, firefighters and search-and-rescue workers could toss these robots through windows and let them scoot around to look for trapped victims or sniff out toxic materials.

① home front：大后方、后方支前活动。

② search-and-rescue workers：搜救人员。

[2] Consequently, their sensors, processing power and physical strength must be distributed among several robots, which must then work in unison.

① physical strength：机械强度。

② in unison：共同，一起。

[3] Something as seemingly inconsequential as the direction of the weave of a rug can dramatically influence their motion, making odometry readings inaccurate, just as a car's odometer would fail to give accurate distances if driven on an ice-covered lake.

① 主句是 Something can dramatically influence their motion，其中 Something 作为主语，can influence 作为谓语，dramatically 作为状语修饰谓语，their motion 作为宾语。

② 定语从句是 as seemingly inconsequential as the direction of the weave of a rug，这里使用了 as…as…结构来强调比较，seemingly inconsequential 作为定语修饰 the direction of the weave of a rug。

③ making odometry readings inaccurate 是现在分词短语，来说明影响的结果。

④ 比较状语从句是 just as a car's odometer would fail to give accurate distances if driven on an ice-covered lake，进一步强调了影响的严重性，其中 a car's odometer 作为主语，would fail 作为谓语，to give accurate distances 作为不定式

短语，if driven on an ice-covered lake 作为条件状语从句。

[4] The team leader—either the home base or a larger robot that deployed the millibots—collects all the information and calculates robot positions using trilateration.

home base：根据地、总部、管理中心。

27.4　Exercise

Translate the following paragraphs

Man has always used augmenters to increase the powers which nature endowed him. A simple stick, made into a lever to extend his reach and enable him to move heavy objects, may have been his first machine. Slingshots helped him kill food animals that were at greater distances. The wheel eased his movement and his burdens over the ground. However, those were simple augmenters for simple tasks. Man has made his world much more complex. With the rapid development of science and technology, man's augmenters have become much more sophisticated. They are designed not only to do the work of man but also to do it in much the same way a man would.

Some simply amplify the muscular power of their human operators. The "walking truck" being constructed is a large device that might better be called "a walking horse". A man sits inside, moving his arms and legs to make the "horse" move its own four legs and carry far greater loads than the operator could. Another man, wearing a movable mechanical and electrical framework, for example, lifts 680kg in 6s.

Other machines link a man to a computer and the computer to a work device. Thanks to the computer, the man does not need to guide the device through each step. The computer "remembers" how to direct the arms of the machine by itself through many work steps.

Still other mimicry machines go where man cannot—into the ocean depths and into nuclear reactors where the radiation would cause a human operator to become sick or to die.

Augmenters that extend man's mental and physical skills over long distances but still require man's remote control are called tele-operators. Nowadays, doctors can use such systems to operate on patients thousands of miles away.

With machines and robots becoming more sophisticated, will they someday take over the world from man? Most scientists don't think so. Instead, they say, robots will take over more of man's heavy work, giving man more time for creative work.

Lesson 28　Introduction to Automobile Engine

28.1　Text

It is well known that the traditional automobile includes engine, chassis, body, and electrical equipment. The engine, often called the heart of an automobile, is used to supply power.

1. Classification of Engine

The engine can be classified in several ways depending on different design features:

(1) The number of cylinders.

Current engine designs include three-cylinder, four-cylinder, six-cylinder, and even eight-cylinder, twelve-cylinder engines. The number of cylinders is one of many factors that determine power and fuel efficiency, but not the only factor. Four-cylinder and six-cylinder engines are the most common at present. A three-cylinder engine could satisfy the needs of energy saving and emission reduction, which is being prioritized by automobile manufacturers.

(2) Cylinder arrangement.

The cylinder arrangement of an engine can be inline, horizontal-opposed, or V-type. More complex designs, such as W8, W12, W16, and W18, have also been used. In-line engines have the cylinders in a line, which is suitable for three-cylinder and four-cylinder engines. As the number of the cylinders increases to six, a V-type design is used to decrease the length of the engine. When the number of cylinders increases further, W-type designs appear. The W-type engine is newly developed, complicated in structure, and expensive in cost.

(3) Fuel type.

The types of fuel are mainly divided into gasoline, diesel, bio-diesel, and ethanol. Gasoline and diesel are the most common automobile fuels and are used all over the world. Diesel fuel created using vegetable oils or animal fats is called bio-diesel. Although ethanol is not widely used as general automobile fuel, it is added to common gasoline as an additive. Many vehicle manufacturers design vehicles that can be powered by ethanol because it is a cost-effective fuel made from renewable resources like corn and sugarcane. With traditional energy prices rising and the world placing more emphasis on pollution reduction and environmental protection, governments, consumers, and vehicle companies have turned to manufacture low-emissions vehicles, so this conventional fuel vehicle could be replaced.

2. Gasoline Engine

The gasoline engine includes two mechanisms: the crank mechanism and the valve mechanism, and five systems: starting system, ignition system, fuel system, lubrication system, and cooling system.

(1) Crank mechanism.

The crank mechanism mainly includes the cylinder head cover, cylinder head, cylinder block, cylinder sleeve, cylinder gasket, oil pan, piston, connecting rod, and crankshaft. It is used to convert the reciprocating motion of the piston into a rotary motion of the crankshaft to drive the vehicle.

The piston is a cylindrical, hollow part made of aluminum alloy. Its reciprocating movements in the cylinder transform the energy of the expanding gases into mechanical energy. The connecting rod connects the piston to the crankshaft. The crankshaft is one of the most important parts of the engine and is held in the crankcase in the cylinder block. The flywheel is a rotating disk located on one end of the crankshaft, used to reduce vibration caused by the power stroke through its inertia and to start the engine when the starter meshes with it.

(2) Valve mechanism.

The valve mechanism is used to open and close the valves at just the right time, achieved by the camshaft rotating. The timing and duration of valve opening is called valve timing. Since different engine speeds different valve timing, some engines are equipped with variable valve timing (VVT) to achieve the best valve timing for high and low speeds.

Most engines have two camshafts. The intake camshaft opens and closes the intake valves, while the exhaust camshaft does the same for the exhaust valves. When the two camshafts are located in the cylinder head, the arrangement is called dual overhead-camshaft (DOHC), which is becoming mainstream among automotive manufacturers. To improve intake and exhaust efficiency, there are generally two intake valves and two exhaust valves per cylinder. The exhaust valve stays open briefly, during which the intake valve is also open. This overlap at the end of the exhaust stroke and the beginning of the intake stoke is called valve overlap.

(3) Starting system.

The starting system mainly consists of a starter, an electromagnetic switch, a control circuit, a starter relay, an ignition switch (a starting switch), a battery, and a starter circuit. The battery, also known as a lead-acid storage battery, is an electrochemical device that produces voltage and delivers current. The key is inserted into the ignition switch and turned to the start position. A small amount of current then passes through the neutral safety switch to a starter relay or starter solenoid which allows high current to flow through the battery cables to the starter motor. [1]The starter motor cranks the engine, causing the piston to move downward and create suction that draws mixtures into the cylinder, where a spark created by the ignition system ignites it. If the compression pressure in the engine is high enough and all this happens at the right time, the engine will start. Once the engine runs, the alternator supplies all the electrical components of the vehicle.

(4) Electronic fuel injection (EFI).

EFI can be divided into three basic subsystems: fuel delivery system, air supplying system, and electronic control system.

The fuel delivery system contains the oil tank, electric fuel pump, fuel filter, oil pressure regulator, oil rail, fuel injector, and so on. Since the injection pressure is kept constant by the oil pressure regulator, the amount of fuel injected depends on injector timing, which is decided by ECU from signals received from various sensors.

[2]The air supply system provides the gasoline engine with clean, metered, fresh air that is compatible with the engine load so that they form a good quality combustible mixture with the gasoline injected in the intake pipe or cylinder. It includes the air cleaner, air flow sensor, throttle valve, intake manifold, and so on. The throttle valve controls how much air enters the cylinder, which is managed by the driver pressing the accelerator pedal.

The electronic control system consists of various engine sensors, ECU, and fuel injector assemblies. The ECU determines precisely how much fuel needs to be delivered by the injector by monitoring the engine sensors.

(5) Ignition system.

The ignition system supplies strong electric sparks to ignite the air-fuel mixture in the combustion chamber at the right time. There are many types of electronic ignition systems, maily classfied into distributor and distributor-less ignition systems. The distributor-less ignition system is computer-controlled and has no moving parts, greatly improving reliability. It may have one ignition coil for per cylinder or one ignition coil for each pair of cylinders.

Coil-on-plug systems have individual ignition coils for each spark plug, installed directly on it, maximizing spark strength and system reliability. This setup is popular for performance, emissions, and maintenance reasons. It eliminates the need for long, bulky spark plug cables, reducing radio frequency interference and misfire problems, and lowering resistance between the ignition coil and spark plug. Consequently, each ignition coil can be smaller, lighter, and use less energy.

(6) Cooling system.

Due to the combustion of fuel with air inside the cylinder, the temperature of the

Introduction to Automobile Engine Lesson 28

engine increases, affecting performance and the lifespan of engine parts. The cooling system maintains efficient operating temperature, preventing overheating and overcooling under all driving conditions. It includes the radiator, fan, coolant pump, water jacket, thermostat, and so on. [3] It involves little and full circulation, which is managed by a thermostat that regulates coolant flow through the radiator.

After starting the engine, it should quickly heat to an efficient operating temperature and maintain it without overheating. The thermostat remains closed initially, and the coolant is recirculated within the engine block and cylinder head. When the temperature reaches a predetermined level, the thermostat begins to open and allows the coolant to flow through the radiator for full circulation. If the thermostat fails and remains closed, the engine overheats because the coolant cannot flow into the radiator. If it remains open, the engine overcools.

(7) Lubrication system.

An engine has many moving parts which develop wear as they move against each other. The engine circulates oil between these parts to prevent metal-to-metal contact and wear. Oiled parts move more easily, reducing friction and minimizing power loss. The lubricant also acts as a coolant and a sealing medium to prevent leakage. A film of lubricant on the cylinder walls helps the piston rings seal, improving the engine's compression pressures.

28.2 Words and Phrases

gasoline	*n.* 汽油
bio-diesel	生物柴油
ethanol	*n.* 酒精，乙醇
cost-effective	经济划算的，成本效益高的
sugarcane	*n.* 甘蔗
cylinder block	气缸体
piston	*n.* 活塞
reciprocating	*adj.* 往复的；摆动的；*n.* 往复运动
flywheel	*n.* 飞轮
inertia	*n.* 惯性
camshaft	*n.* 凸轮轴
electrochemical	*adj.* 电气化学的
starter solenoid	起动机电磁线圈
air cleaner	空气滤清器

air flow sensor	空气流量计
throttle valve	节气门
fuel injector	喷油器
ignition coil	点火线圈
spark plug	火花塞
coil-on-plug	独立点火系统
thermostat	n. 节温器

28.3　Complex Sentence Analysis

［1］The starter motor cranks the engine causing the piston to move downward and create suction that draws mixtures into the cylinder, where a spark created by the ignition system ignites it.

① that 引导定语从句，修饰前面的 suction。
② where 作为关系副词引导定语从句，修饰前面的 cylinder。
③ created by the ignition system 为过去分词短语，放在名词后作后置定语，修饰前面的 a spark。

［2］The air supply system provides the gasoline engine with clean, metered, fresh air that is compatible with the engine load so that they form a good quality combustible mixture with the gasoline injected in the intake pipe or cylinder.

① that is compatible with the engine load 为定语从句，与 clean, metered, fresh 一起修饰 air。
② so that 引导目的状语从句。
③ injected in the intake pipe or cylinder 为过去分词短语作后置定语，修饰 gasoline。

［3］It involves little and full circulation, which is managed by a thermostat that regulates coolant flow through the radiator.

本句为双重定语从句，which is managed by a thermostat 修饰 little and full circulation，that regulates coolant flow through the radiator 修饰 thermostat。

28.4　Exercise

Translate the following paragraphs

Fuel stratified injection (FSI) is a proprietary direct fuel injection system developed and used by Volkswagen AG and its luxury subsidiary Audi.

As with all direct-injection systems, FSI increases the engine's power output while reducing fuel consumption by as much as 15%. This is largely due to the stratified charge principle at part load. In this mode, the engine only requires a fuel-air mixture capable of immediate ignition around the spark plug. The rest of the combustion chamber is filled with a leaner mixture, containing excess air.

Turbo fuel stratified injection (TFSI) is a trademark of the Volkswagen Group for a type of forced-aspiration (turbo) engine where the fuel is pressure-injected directly into the combustion chamber to create a stratified charge.

Lesson 29　Introduction to Automobile Chassis

29.1　Text

The chassis is composed of power train system, steering system, suspension system, and braking system, which supports the body, accept the driving force generated by the engine, and ensure the vehicle drives normally.

1. Power Train System

[1] The power train system transfers power from the engine to the driving wheels, which mainly includes the clutch that is only required with the manual transmission to temporarily disconnect the engine from the driving wheels, transmission, final drive, differential, and driving wheels. The location of the engine and driving wheels determines whether the vehicle is classified as FF, FR, RR, MR, or nWD.

(1) Clutch.

When the clutch pedal is depressed, it disengages the engine from the transmission. When the clutch pedal is released, the pressure plate forces the clutch disc against the flywheel.

(2) Transmission.

The transmission is a key device that makes better use of engine power and torque. We can choose more speed and less torque (higher gear) or less speed and more torque (lower gear) to suit different driving conditions. There are mainly three types of transmissions: manual transmission (MT), automatic transmission (AT), and continuously variable transmission (CVT).

① MT.

The MT can change the gear meshing position by adjusting the gear-shifting position by hand. MT usually has five forward gears, while some vehicles now have as many as six or seven gears. The more forward gears, the smoother the shifting, and the more fuel-efficient the vehicle becomes.

② AT.

The AT is the most complicated mechanical component in the vehicle. It consists of mechanical system, hydraulic system, and electronic control system, all working together in perfect harmony.

The AT uses a planetary gear mechanism for shifting. It automatically shifts according to the degree of the accelerator pedal and the speed of the vehicle. The driver only needs to operate the accelerator pedal to control the speed.

2. Steering System

The steering system provides the driver with a means to control the vehicle's direction as it moves. As the heart of the steering system, the steering box has many types, but two of them are most popular: the rack-and-pinion steering box used in many passenger vehicles, and the recirculating ball steering box used in most commercial vehicles, especially heavy ones. The rack-and-pinion steering box is small, lightweight, and easy to service. It gives more feedback and road feel to the driver. The recirculating ball steering box is durable, with good steering response and road feel for the driver.

Power steering system adds a series of assisting mechanism to reduce driver effort. Depending on the type of power, it can be mainly classified into hydraulic power-assisted steering system (HPAS) and electric power-assisted steering system (EPAS).

(1) HPAS.

HPAS consists of an oil reservoir, a steering pump, a steering control valve, a steering cylinder, and corresponding pipes. The steering control valve provides oil pressure which corresponds to the rotary motion of the steering wheel for the steering cylinder. The steering cylinder converts the applied oil pressure into an assisting force which acts on the steering box and intensifies the steering force exerted by the driver. When the steering wheel is centered, the steering control valve is not actuated, and the oil delivered by the steering pump flows back to the steering oil reservoir. When the steering wheel rotates, the steering control valve opens or closes accordingly. The oil on one side will directly flow back to the steering oil reservoir, while the oil on the other side continues to flow into the hydraulic cylinder. This creates a pressure difference on both sides of the piston, generating auxiliary force, which makes steering easier.

(2) EPAS.

EPAS uses an electric motor to provide directional control of the vehicle. It mainly contains torque sensor, speed sensor, ECUs, and electric motor. The electric power steering system can adjust the amount of power according to the speed, making the steering wheel lighter at low speeds and more stable at high speeds. EPAS is replacing HPAS and is becoming mainstream among automotive manufacturers. Since EPAS does not require engine power to operate, a vehicle equipped with EPAS may reduce fuel consumption, thus improving the automobile's power performance.

3. Suspension System

The suspension system includes suspension, vehicle frame, axle, and wheels. It is used to flexibly connect the wheels to the body, transfer power from the engine via the power train system, and quickly dampen spring oscillations.

(1) Suspension.

The suspension covers the arrangement used to connect the wheels to the body, which includes spring elements, shock absorbers, and guide mechanisms. It can be classified into independent suspension and dependent suspension. For the independent suspension, each wheel mounts separately to the frame and has its own individual spring and shock absorber. Thus, the wheels act independently of one another, improving driving comfort.

There are three basic types of springs in automobile suspension: coil springs, leaf springs, and torsion bars. Coil springs are the most common type used on almost all vehicles. When a spring is compressed and then released, it will oscillate for a period before coming to rest. Applied to a vehicle, this action causes an uncomfortable driving. Therefore, vehicles are equipped with shock absorbers to absorb the energy stored in the spring and reduce the bouncing time. Hydraulic shock absorbers are widely used in automobile suspension systems.

(2) Four-wheel alignment.

The four-wheel alignment is necessary for today's vehicle designs. Correct alignment is vital for vehicle control, not only for safety but also for comfort while driving. A four-wheel aligner is a precise measuring instrument used to measure wheel alignment parameters and compare them with the specifications provided by the vehicle manufacturer. It also provides instructions for performing corresponding adjustments to achieve the best steering performance and reduce tire wear.

4. Braking System

The braking system is the most important safety system in an automobile. Its function is to stop the vehicle within the shortest possible distance by converting the vehicle's kinetic energy into heat energy, which is dissipated into the atmosphere. Typically, each automobile has two completely independent braking systems according to different functions: the service braking system, which is normally foot-operated, and the parking braking system, which is hand-operated.

With the constant development of automobile electrical technology, the parking braking system has gradually evolved from a primary mechanical type to an electrical type. [2] Providing an additional level of safety in today's vehicles, electric parking braking system also improves driver's convenience by ensuring driver-assist functions, including automatic brake release when moving off and a hill-hold function for incline starts.

A service braking system consists of an energy-supplying device, a control device, a transmission device, and a brake. A brake inhibits motion and is a decisive part of a vehicle. It absorbs energy from the moving part and slows down the vehicle using friction. Each wheel has a brake assembly, which can be either the drum type or the disc type, hydraulically or pneumatically operated when the driver depresses the brake pedal.

(1) Drum brake.

A drum brake has two shoes, anchored to a stationary back-plate, which are internally expanded to contact the drum by hydraulic cylinders. When the brake pedal is depressed, the pistons in the master cylinder move, forcing brake fluid through the brake lines to the wheel cylinder. This causes the brake shoes to contact the brake drum, applying friction to slow or stop the wheel. When the brake pedal is released, the return springs contract and pull the shoes away from the braking surface, while the movement of the wheel cylinder pistons forces fluid back to the master cylinder to replenish the reservoir.

(2) Disc brake.

Hydraulically actuated disc brakes are the most commonly used forms of brake for passenger vehicles. There are two kinds of disc brakes currently used in modern automobiles: disc brakes with a fixed caliper and disc brakes with a floating caliper.

The main disadvantage of drum brakes is that the friction area is almost entirely covered by a lining, so most heat must be conducted through the drum to reach the outside air to cool. Because of being exposed to the air, disc brakes dissipate heat easier, giving them greater resistance to fade (reduction in brake efficiency due to heat) than drum brakes. This means disc brakes can operate continuously for longer period and provide better gradual braking efficiency. There are some other advantages of disc brakes, such as equal wear of the inboard and outboard brake pads and relatively constant brake performance with lower susceptibility to fade. However, they also have disadvantages, such as shorter brake pad life when used on heavy-duty commercial vehicles, higher acquisition and operating costs, and a tendency to cause brake noise. Drum brakes have more parts than disc brakes and are harder to service, but they are less expensive to manufacture and can easily incorporate an emergency brake mechanism. The effective brake friction area of drum brakes is larger than that of disc brakes, giving drum brakes higher braking efficiency, making them necessary for heavy-duty commercial vehicles. Disc brakes have been used in most passenger vehicles and are now being adopted in commercial vehicles as well.

29.2　Words and Phrases

chassis	*n.* 底盘
clutch	*n.* 离合器，离合器踏板
temporarily	*adv.* 临时地，暂时地
differential	*n.* 差速器
planetary gear mechanism	行星齿轮机构
steering control valve	转向控制阀
steering cylinder	转向液压缸
suspension	*n.* 悬架装置
shock absorber	减振器
coil spring	螺旋弹簧

leaf spring	钢板弹簧
torsion bar	扭杆弹簧
four wheel alignment	四轮定位
service braking system	行车制动系统
drum brake	鼓式制动器
disc brake	盘式制动器

29.3　Complex Sentence Analysis

［1］ The power train system transfers power from the engine to the driving wheels, which mainly includes the clutch that is only required with the manual transmission to temporarily disconnect the engine from the driving wheels, transmission, final drive, differential, and driving wheels.

本句为双重定语从句。which mainly includes the clutch, transmission, final drive, differential, and driving wheels 修饰 the power train system。that is only required with the manual transmission to temporarily disconnect the engine from the driving wheels 修饰 clutch。

［2］ Providing an additional level of safety in today's vehicles, electric parking braking system also improves driver convenience by ensuring driver-assist functions, including automatic brake release when moving off and a hill-hold function for incline starts.

① Providing an additional level of safety in today's vehicles 为动名词作前置定语，表示事物的属性、用途等。
② by ensuring driver-assist functions 为方式状语。
③ including automatic brake release 为介词短语作后置定语。
④ when moving off and a hill-hold function for incline starts 为时间状语。

29.4　Exercise

Translate the following paragraph

The antilock braking system (ABS) enhances safety and convenience by modulating the hydraulic pressure in the braking system to prevent the brakes from locking and the tires from skidding on slippery pavement or during a panic stop. ABS modulates brake application force several times per second to maintain a controlled amount of slip; all systems achieve this in essentially the same way. One or more speed sensors generate an alternating current signal whose frequency increases with the wheels rotational speed. An electronic control unit continuously monitors these signals, and if the frequency of a signal drops too rapidly, indicating that a wheel is about to lock, the control unit instructs a modulating device to reduce hydraulic pressure to the brake at the affected wheel. When sensor signals indicate the wheel is rotating normally again, the control unit allows increased hydraulic pressure to the brake. This release-apply cycle occurs several times per second, simulating a rapid pumping action similar to what a driver might do, but at a much faster rate. In addition to their basic operation, ABSs have two other things in common. They do not activate until the brakes are applied with enough force to lock or nearly lock a wheel. At all other times, ABS remains ready to function but does not interfere with normal braking. If ABS fails in any way, the brakes continue to operate without anti-lock capability. A warning light on the instrument panel alerts the driver when there is a problem with the ABS.

Lesson 30 New Energy Vehicle

30.1 Text

Speeding up the development of new energy vehicles is one of the most important ways for the world to move towards carbon neutrality. With the development and improvement of new energy vehicle technology, there are also many new opportunities for the industry. In China, "New Energy Vehicle (NEV) Industry Development Plan (2021—2035)" aims to promote and stimulate high-quality, sustainable development of China's new energy automotive industry. This plan, released by the State Council, listed five strategic tasks to improve technology innovation capacity, build new-type industry ecosystems, advance industrial integration and development, perfect the infrastructure system, and deepen opening-up and cooperation. By 2035, battery electric vehicles are likely to become the mainstream in the sales of new ones, while those used in public transportation will be exclusively electrified.

Electric vehicles reduce dependence on petroleum and tap into a source of electricity that is often domestic and relatively inexpensive. The major driving factors for the electric vehicle market include environmental pressure, energy security, technological advancement, and support from government policies. On the consumption front, preferential policies have encouragedconsumers to choose new energy vehicles over fossil-fueled ones, facilitating the expansion of the NEV market and powering firms along the industrial chain to step up innovation.

"The Administrative Provisions on Production Enterprises of New Energy Vehicles and Access of Products (Amended Draft for Comments)", issued by the Ministry of Industry and Information Technology on 12 August 2016, records an updated definition of NEVs and stricter standards for technical criteria of NEVs. In accordance with the measures, NEVs are vehicles which adopt innovative power systems and are mainly or completely driven by new energy, including plug-in hybrid-electric vehicle (PHEV), battery electric vehicle (BEV), and fuel cell electric vehicle (FCEV). Compared with traditional vehicles, NEV adopts unconventional fuel as power source or uses conventional automotive fuel with innovative engine installations. With advanced technical principles, new technologies, and new structures, NEV is integrated with advanced technologies in power control and driving.

New Energy Vehicle Lesson 30

1. PHEV

A PHEV is a hybrid with high-capacity batteries that can be charged by plugging into an electrical outlet or charging station. It can store enough electricity to significantly reduce petroleum use under typical driving conditions.

There are two basic plug-in hybrid configurations. A series plug-in hybrid electric vehicle, also called an extended range electric vehicle (EREV), uses only the electric motor to turn the wheels. The gasoline engine only generates electricity. It can run solely on electricity until the battery runs down, at which point the gasoline engine generates electricity to power the electric motor. For short trips, these vehicles might use no gasoline at all. For the parallel or blended plug-in hybrid electric vehicle, both the engine and electric motor are connected to the wheels and propel the vehicle under most driving conditions. Drive train Schematic Diagram of PHEV is shown in Figure 30.1. Electric-only operation usually occurs only at low speeds.

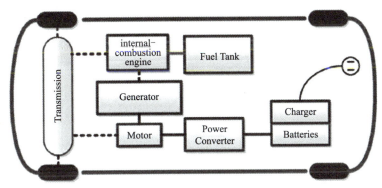

Figure 30.1 Drive Train Schematic Diagram of PHEV

Some PHEVs have higher-capacity batteries and can go further on electricity than others. PHEV fuel economy can be sensitive to driving style, driving conditions, and accessory use.

2. BEV

A BEV is more frequently called EV, is fully electric vehicle with rechargeable batteries and no gasoline engine. All energy to run the vehicle comes from the battery pack, which is recharged from the grid. BEVs are zero-emission vehicles, as they do not generate any harmful tailpipe emissions or air pollution hazards caused by traditional gasoline-powered vehicles. The batteries in BEVs must be recharged periodically. BEVs most commonly charge from the national power grid, through a domestic plug charging point, public charging stations, or a bespoke commercial outlet. Although charging time is limited by the capacity of the grid connection, BEVs are normally designed to be fully charged overnight. The charging infrastructure required to support electric vehicles has many technical issues, requiring the development of low-cost, reliable mobility solutions.

3. FCEV

An FCEV has the potential to significantly reduce dependence on foreign oil and lower harmful emissions that contribute to climate change. FCEVs run on hydrogen gas rather than gasoline and emit no harmful tailpipe emissions. Several challenges must be overcome for them to be competitive with conventional vehicles, but their potential benefits are substantial. FCEVs look like conventional vehicles but use cutting-edge technologies. The heart of an FCEV is the fuel cell stack, which converts hydrogen gas stored onboard with oxygen from the air into electricity to power the vehicle's electric motor. Major components of FCEV are illustrated in Figure 30.2.

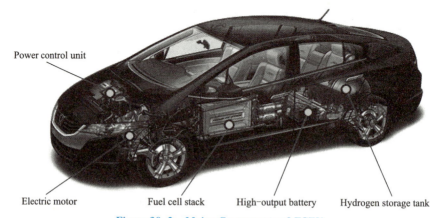

Figure 30.2　Major Components of FCEV

Fuel cells are electrochemical devices that convert chemical energy into electrical energy without combustion. [1]Unlike a battery, a fuel cell can continuously produce electricity as long as the fuel is supplied to it. The storage of fuel, especially hydrogen, is problematic as hydrogen is inflammable and can easily escape from pressure vessels. Depending on the material that the pressure vessel is made from, some materials can experience hydrogen cracking, adding to fatigue and long-term durability issues. Hydrogen is an attractive fuel for vehicles powered by fuel cells, but its transport and storage present challenges. [2]The technical issues in using hydrogen as a fuel are that, although hydrogen can be transported by pipelines, hydrogen tends to leak and can embrittle some metals used for pipelines and valves.

Unlike BEVs, FCEVs can be quickly refueled, similar to conventional vehicles, in only a few minutes, and can drive more quietly, smoothly, and efficiently than other kinds of vehicles. FCEVs are considered as the next generation of electric vehicles.

China's new energy vehicle (NEV) sector has turbocharged growth in recent years, getting a head start in the global race to electrify vehicles and fostering competitive edges in its domestic auto brands. With policy support and market-driven innovation, China has put in place a relatively complete NEV industrial chain, covering batteries, electrical motors, electronic controls, and vehicle manufacturing and sales. Amid the global wave of electrification and intelligent transformation, China's carmakers must step up technological innovation and make vehicle manufacturing more digitalized and intelligent. China opens its arms wide to welcome foreign NEV makers, who have found it necessary to cooperate with their Chinese peers or to participate in China's NEV supply chains to increase their competitiveness in the global market.

30.2 Words and Phrases

carbon neutrality	碳中和
sustainable	adj. 可持续性的
battery swap	换电池
tap into	使发挥最大功效
petroleum	n. 石油
registration tax	登记税
value-added tax	增值税
utilization	n. 利用
subsidise	vt. 给……津贴
generalized	adj. 广泛的；普遍的
post subsidy	额外补贴
periodically	adv. 周期性地，定期地
bespoke	adj. 订做的
substantial	adj. 重大的
sensitive	adj. 敏感的
competitive	adj. 竞争的，有竞争力的
fuel cell stack	燃料电池堆
inflammable	adj. 易燃的
embrittle	v. 使变脆

30.3 Complex Sentence Analysis

[1] Unlike a battery, a fuel cell can continuously produce electricity as long as the fuel is supplied to it.

as long as 引导条件状语从句，意为"只要……"。

[2] The technical issues in using hydrogen as a fuel are that, although hydrogen can be transported by pipelines, hydrogen tends to leak and can embrittle some metals used for pipelines and valves.

① in using hydrogen as a fuel 为介词短语作后置定语。
② although 引导让步状语从句。

30.4 Exercise

Translate the following paragraph

Regenerative braking is a unique technique that is used in BEVs to capture the kinetic energy of a vehicle, which would otherwise be wasted during deceleration or braking. This technique converts the vehicle's kinetic energy back into electrical energy during braking (deceleration or downhill running). The converted electrical energy is stored in energy storage devices such as batteries, ultracapacitors, and ultrahigh-speed flywheels, extending the driving range by up to 10%. Integrating various systems to optimize efficiency, regenerative braking is a vital component of that overall equation. Besides minimizing energy loss and extending electric range, regenerative braking systems lengthen brake life due to their low wear and tear characteristics. This technology makes green vehicles more appealing to consumers.

Lesson 31　Autonomous Vehicle

31.1　Text

An autonomous vehicle is a vehicle capable of sensing its environment and operating without human involvement, also known as a self-driving vehicle, a driverless vehicle, or a robotic vehicle. A human passenger is not required to take control of the vehicle at any time, nor is a human passenger required to be present in the vehicle at all. [1] An autonomous vehicle can go anywhere a traditional vehicle goes and do everything that an experienced human driver does. A fully autonomous vehicle would be self-aware and capable of making its own choices. Autonomous vehicles are becoming standard as these technologies continue to mature.

Driving automation have been classified into six levels (from level 0 to level 5) by the national standard "Taxonomy of Driving Automation for Vehicles" as follows:

Level 0 driving automation (emergency assistance) is not non-driving automation. Level 0 systems can perceive the environment and provide information or briefly intervene in vehicle control to help drivers avoid danger. Level 1 driving automation (partial driver assistance) systems and level 2 driving automation (combined driver assistance) systems work with the driver to execute all dynamic driving tasks, supervising the system's behavior and responding or operating appropriately.

For level 3 driving automation (conditionally automated driving), the user takes over suitably. In level 4 driving automation (highly automated driving), when the system issues an intervention request, the user may not need to respond, as the system can automatically reach a minimal risk condition.

[2] Level 5 driving automation involves fully automated driving, whereby the system issues an intervention request, the user may not need to respond, as the system can automatically reach a minimal risk condition. Level 5 driving automation has no design operational domain restrictions in environments where the vehicle can operate (excluding commercial and regulatory limitations).

The application of 5G will enable level 5 driving automation, allowing vehicles not only to communicate with each other but also with traffic signals, traffic signs, and even the road itself.

1. Working Principle of Autonomous Vehicles

Autonomous vehicles rely on sensors, actuators, complex algorithms, machine learning systems, and powerful processors to execute software. They create and maintain a map of their surroundings based on a variety of sensors situated in different parts of the vehicle. Radar sensors monitor the position of nearby vehicles. Video cameras detect traffic lights, read road signs, track other vehicles, and look for pedestrians. Lidar sensors (light detection and ranging) bounce pulses of light off the vehicle's surroundings to measure distances, detect road edges, and identify lane markings. Ultrasonic sensors in the wheels detect curbs and other vehicles when parking. Sophisticated software then processes all the sensory input, plots a path, and sends instructions to the vehicle's actuators, which control acceleration, braking, and steering. Hard-coded rules, obstacle avoidance algorithms, predictive modeling, and object recognition help the software follow traffic rules and navigate obstacles.

(1) Adaptive cruise control (ACC) system.

One aspect of the technology used in autonomous vehicles is ACC system. This system can adjust the vehicle's speed automatically to ensure it maintains a safe distance from the vehicles in front. This function relies on information obtained using sensors on the vehicle and allows the vehicle to perform tasks such as braking when it senses it is approaching vehicles ahead. This information is processed, and the appropriate instructions are sent to actuators in the vehicle, which control responsive actions such as steering, acceleration, and braking. Highly automated vehicles with fully automated speed control can respond to signals from traffic lights and other non-vehicular activities.

(2) Advanced driving assistance system (ADAS).

ADAS is a system designed to help drivers in the driving process. When designed with a safe human-machine interface, they should increase vehicle safety and road safety in general.

The paramount goal of ADAS is to assist the driver with the safety aspects for themselves and for other mobile and pedestrian traffic participants. ADAS also aims to enhance driving comfort and improve the economic and environmental balance.

Most road accidents occur due to human errors. ADASs are systems developed to automate, adapt, and enhance vehicle systems for safety and better driving. The automated system provided by ADAS is proven to reduce road fatalities by minimizing human errors. Safety features are designed to avoid collisions and accidents by offering technologies that alert the driver to potential problems or avoid collisions by implementing safeguards and taking over control. Adaptive features may automate lighting, provide ACC, collision avoidance, pedestrian crash avoidance mitigation (PCAM), incorporate traffic warnings, connect to smart phones, alert the driver to other vehicles or dangers, offer a lane departure warning system, automatic lane centering, or show what is in blind spots.

Many ADASs are on the cutting edge of emerging automotive technologies, and the jury is actually still out on some of them. Some of these systems will have the staying power to stick around, and you can expect to see at least a few of them in your next vehicle. Others may fizzle and disappear or be replaced by better implementations of the same basic idea.

With the deepening application of voice recognition technology, automotive engineers are also seeking ways to better integrate it into automotive systems. When developers regard voice recognition as the core of a complex human-machine interface, the field of control technology is bound to undergo transformative changes.

2. Advantages of Autonomous Vehicles

Autonomous vehicle technology may provide certain advantages compared to human-driven vehicles.

One potential advantage is increased safety on the road. [3] Vehicle crashes cause many deaths every year, and automated vehicles could potentially decrease the number of casualties as the software used in them is likely to make fewer errors in comparison to humans. A decrease in accidents could also reduce traffic congestion, which is a further potential advantage of autonomous vehicles. Autonomous driving can achieve this by removing human behaviors that cause blockages on the road, specifically stop-and-go traffic. Another possible advantage of automated driving is that people who are unable to drive due to factors like age and disabilities could use automated vehicles as more convenient transport systems. Additional advantages include the elimination of driving fatigue and the ability to sleep during overnight journeys.

3. Obstacles to Adoption of Autonomous Vehicles

The technology of fully autonomous vehicles is currently being tested, but we are still far from making them available to the general public. Although the autonomous vehicle has come a long way, it is not yet a common mode of transportation, with various obstacles to overcome. In particular, some technologies required to build a functional self-driving vehicle are very expensive, making the final cost prohibitive for the general public. Radar and Lidar work for prototypes, but if mass production is achieved, their signals and frequencies might interfere with one another. Anomalous conditions such as snow, debris, or oil may represent significant challenges whenever they cover lane markings and dividers. Also, it is still early to trust AI to be smart enough to make consistent decisions or split-second judgement calls in life-or-death scenarios, for example, when a pedestrian suddenly traverses the road after a steep turn.

31.2 Words and Phrases

autonomous   *adj.* 自治的，独立自主的，自律的，自制的

algorithm	*n*. 运算法则
pedestrian	*n*. 行人，步行者
lidar sensor	激光雷达传感器
curb	*n*. 路缘（俗称马路牙子）
ultrasonic sensor	超声波雷达传感器
traffic congestion	交通拥堵
driving fatigue	驾驶疲劳

31.3 Complex Sentence Analysis

[1] An autonomous vehicle can go anywhere a traditional vehicle goes and do everything that an experienced human driver does.

① anywhere 引导地点状语从句。

② that 引导定语从句。

[2] Level 5 driving automation involves fully automated driving, whereby the system issues an intervention request…

whereby 引导定语从句，作关系副词，同 by which。

[3] Vehicle crashes cause many deaths every year, and automated vehicles could potentially decrease the number of casualties as the software used in them is likely to make fewer errors in comparison to humans.

① as 引导原因状语从句。

② in comparison to…：与……相比。

31.4 Exercise

Translate the following paragraph

The earliest edge of the autonomous driving frontier was the evolution of multiple driver safety features that are now standard in many new vehicles. For example, a lane departure warning system alerts drivers if the vehicle seems to be leaving a particular space on a multilane road. Parking assist features, automated braking, and other features also apply. Each one has its own specialized function and controls a given task. What these systems have in common is that, although they promote the idea of autonomous vehicle design, they still do not approach what some would call a self-driving vehicle. The human driver is still in control and needs to be in control, and the warnings and alerts is utilized to make better driving decisions. These forms of partial driving automation that still require a human driver pertain to level 0 to level 2. Only when the automated system is in charge of monitoring the driving environment and human intervention is purely optional can a vehicle be defined as "autonomous" (level 3 to level 5). As we progress toward autonomous vehicle design, automated systems are becoming more common.

Lesson 32 How to Write a Scientific Paper

32.1 Text

A scientific paper is a written and published report describing original research results. This definition must be qualified, however, by noting that a scientific paper must be written and published in a certain way.

1. Title

In preparing a title for a scientific paper, the author would do well to remember one salient fact that the title will be read by thousands of people. Perhaps few people, if any, will read the entire paper, but many people will read the title, either in the original journal or in one of the secondary (indexing and abstracting) services. Therefore, all words in the title should be selected with great care, and their association with one another must be carefully managed.

We can define a good title with the fewest words that adequately describe the contents of the scientific paper.

The title of a scientific paper is like a label. Because it is not a sentence, with the usual subject, verb, object arrangement, it is really simpler than a sentence (or, at least, usually shorter), but the order of the words becomes even more important.

The meaning and order of the words in the title are of importance to the potential reader who reads the title in the journal table of contents. These considerations are equally important to all potential users of the literature, including those (probably a majority) who become aware of the scientific paper via secondary sources. Thus, the title should be useful as a label accompanying the scientific paper itself, and it also should be in a form suitable for the machine-indexing systems used by the Engineering Index (EI), Science Citation Index (SCI), and others. Most of the indexing and abstracting services are geared to "key word" systems. Therefore, it is fundamentally important that the author provide the right "keys" to the scientific paper when labeling it. That is, the terms in the title should be limited to those words that highlight the significant contents of the scientific paper in terms that are both understandable and retrievable.

2. Abstract

An abstract is a concise and precise summary of the scientific paper. The role of the abstract is not to evaluate or explain, but rather to describe the scientific paper (dissertation). [1] The abstract should include a brief but precise statement of the problem or issue, a description of the research method and design, the major findings and their significance, and the principal conclusion. [2] The abstract should contain the most important words referring to the methods and contents of the scientific paper. These facilitate access to the abstract by computer research and enable readers to identify the basic content of the scientific paper quickly and accurately, to determine its relevance to their interests, and thus to decide whether they need to read the scientific paper in its entirety.

An abstract should be written in complete sentences, rather than in phrases and expressions. Generally, an abstract for a short scientific paper is limited to a maximum of 200—250 words. The abstract should be designed to define clearly what is dealt with in the paper. Many people will read the abstract, either in the original journal or in the EI, SCI, or one of the other secondary publications.

The abstract should never give any information or conclusion that is not stated in the scientific paper. References to the literature must not be cited in the abstract (except in rare instances, such as modification of a previously published method). The abstract is neither numbered nor counted as a page.

3. Introduction

Experienced writers prepare their title and abstract after the scientific paper is written, even though by placement these elements come first. You should, however, have a provisional title and an outline of the scientific paper that you propose to write in mind (if not on paper). You should also consider the level of the audience you are writing for, so that you will have a basis for determining which terms and procedures need definitions or descriptions and which do not.

The first section of the text proper should be the introduction. The purpose of the introduction should supply sufficient background information and the design idea to allow the reader to properly understand and evaluate the results of the present study without needing to refer to previous publications on the topic. The introduction should also provide the rationale for the present study. Above all, you should state briefly and clearly your purpose in writing the scientific paper. Choose references carefully to provide the most important background information.

Suggested rules for a good introduction are as follows:

(1) It should present first, with all possible clarity, the nature and scope of the problem investigated.

(2) It should review the pertinent literature to orient the reader.

(3) It should state the method of the investigation. If deemed necessary, the reasons for the choice of a particular method should be stated.

(4) It should state the principal results of the investigation.

(5) It should state the principal conclusions suggested by the results.

4. Materials and Methods

In the "Materials and Methods" section, you must give full details. Most of this section should be written in the past tense. The main purpose is to describe the experimental design and provide enough details for a competent worker to repeat the experiments. Many readers will skip this section as they already know the general methods from the introduction, but careful writing is critical. The scientific method requires that your results are reproducible; thus, you must provide the basis for others to repeat the experiments.

A good reviewer will read the "Materials and Methods" carefully during peer review. If there is doubt about the reproducibility of your experiments, the reviewer may recommend rejection, regardless of the results.

Materials include exact technique specifications, qualities, source, and method of preparation. List pertinent chemical and physical properties of specimens or reagents used.

Method should be presented chronologically, though related methods should be described together. If your method is new, you should provide all necessary details. If a method has been published in a standard journal, only the literature reference will be needed.

5. Results

The core of the scientific paper is the data. This part of the scientific paper is called the "Results" section.

There are usually two components of the "Results" section. Firstly, you should provide an overall description of the experiments, giving a "big picture" without repeating the experimental details previously provided in the "Materials and Methods" section. Secondly, you should present the data.

Of course, it is not quite easy. A simple transfer of data from the laboratory notebook to the manuscript will hardly do. Most importantly, in the manuscript, you should present representative data rather than endlessly repetitive data.

The "Results" section need to be clearly and simply stated because it is the results that comprise the new knowledge you are contributing to the world. The earlier sections of the scientific paper ("Introduction" and "Materials and Methods") are designed to explain why and how you obtained the results; while the later section ("Discussion") of the scientific paper is designed to interpret what they mean. Obviously, therefore, the whole scientific paper must stand or fall on the basis of the results. Thus, the results must be presented with clarity.

6. Discussion

The "Discussion" section is harder to define than the other sections, making it usually the hardest section to write. Whether you know it or not, many scientific papers are rejected by journal editors because of a faulty "Discussion", even if the data are both valid and interesting. More often, the true meaning of the data may be completely obscured by the interpretation presented in the "Discussion" section, leading to rejection.

Essential features of a good "Discussion" are as follows:

(1) Try to present the principles, relationships, and generalizations shown in the results. In a good "Discussion" section, you discuss, rather than recapitulate the results.

(2) Point out any exceptions or lack of correlation and define unsettled points. Never risk covering up or fudging data that do not quite fit.

(3) Show how your results and interpretations agree or contrast with previously published work.

(4) Discuss the theoretical implications of your work, as well as any possible practical applications.

(5) State your conclusions as clearly as possible.

(6) Summarize your evidence for each conclusion.

In showing the relationships among observed facts, you do not need to reach cosmic conclusions. Seldom will you illuminate the whole truth; more often, you can shine a spotlight on one area of truth. Your area of truth should be supported by your data; if you extrapolate beyond what your data show, you may appear foolish, casting even your data-supported conclusions into doubt.

When describing the meaning of your bit of truth, do it simply. The simplest statements evoke the most wisdom; verbose language and fancy technical words often convey shallow thoughts.

32.2　Words and Phrases

salient	*adj.* 明显的
adequately	*adv.* 充分地
citation	*n.* 引用
retrievable	*adj.* 可检索的
dissertation	*n.* （学位）论文，专题
relevance	*n.* 关联，适用
preliminary	*adj.* 预备的，初步的
provisional	*adj.* 暂定的，假定的
rationale	*n.* 基本原理，理论基础，原理的阐述
above all	尤其是，最重要的是，首先是
pertinent	*adj.* 有关的，相干的，中肯的
cornerstone	*n.* 基石；基础；（建筑）隅石
manuscript	*n.* 手稿，原稿

awe-inspiring	令人敬畏的，令人鼓舞的
specification	n. 详述，[常 pl.] 规格，说明书，规范，明细表
specimen	n. 试样，样品，标本
reagent	n. 反应物，反应力，试剂
chronological	adj. 按时间顺序排列的，按年代顺序排列的
ingredient	n. 成分，要素，因素，原料
obscure	adj. 模糊的，含糊的，晦涩的，暗的，朦胧的；v. 使……黑暗，使不明显
injunction	n. 命令，指令
heed	v. & n. 注意，留心
recapitulate	v. 扼要重述，概括，重现，再演
unsettled	adj. 不稳定的，不安定的，未解决的，混乱的
correlation	n. 关联，相关性，相互关系
cover up	包裹，隐藏，掩盖
fudge	n. 捏造，梦话，胡话，空话；v. 蒙混，逃避责任；vt. 粗制滥造，捏造推诿；int. 胡说八道
implication	n. 牵连，受牵累，暗示，隐含，意义，本质
cosmic	adj. 宇宙的，全世界的，广大无边的
illuminate	v. 照明，照亮，阐明，说明，着凉，使光辉灿烂，以灯火装饰（街道等）；vi. 照亮
spotlight	n. 聚光灯，点光源，公众注意中心
buttress	n. 支持物，支柱；v. 支持，加强，扶住
extrapolate	n. 推断，外推，外插
evoke	v. 唤起，引起，移送
verbose	adj. 冗长的，累赘的，喋喋不休的
fancy	adj. 奇特的，美妙的，漂亮的；v. & n. 设想，嗜好，爱好

32.3 Complex Sentence Analysis

[1] The abstract should include a brief but precise statement of the problem or issue, a description of the research method and design, the major findings and their significance, and the principal conclusion.

a brief but precise statement of the problem or issue、a description of the research method and design、the major findings and their significance 及 the principal conclusion 均为 include 的宾语。

[2] The abstract should contain the most important words referring to the methods and contends of the scientific paper. These facilitate access to the abstract by computer research and enable readers to identify the basic contents of the

scientific paper quickly and accurately, to determine its relevance to their interests, and thus to decide whether they need to read the scientific paper in its entirety.

these facilitate access to…, and enable readers to…entirety 为并列句，facilitate 与 enable 为两个并列的谓语；to identify…、to determine…及 to decide…均为 reader 的宾语补足语。

32.4　Exercise

Translate the following paragraph

A new machining model has been developed for single-point diamond turning of brittle materials. Experiments using the interrupted cutting method allow model parameters to be determined, providing a quantitative method for determinig the machineability of a material with respect to the rake angle, tool nose radius, and machining environment. The model uses two parameters, the critical depth of cut and the subsurface damage depth, to characterize the ductile-regime material removal process. Also included in the model is a parameter used to set the process limit, defined as the maximum feed rate. Machining experiments have verified the model and allow for the determination of optimum machining conditions.

附录 A 关于科技英语翻译

一、翻译

翻译就是将一种语言所表达的意思用另一种语言表达出来，是一项非常复杂的语言转换活动。

翻译是一门艺术，艺术讲究美感，所以翻译需在忠实于原文（准确）的基础上发挥灵感并创造才能。

翻译也是一门科学，翻译学是研究各种语言体系之间相互关系的科学。现代人大量吸收信息，充实自我。作为信息的载体，语言已冲破国界的限制。信息技术飞速发展，极大地促进了信息交流，而语言障碍成为首要问题，所以翻译不仅要有追求真理的科学态度，而且要逐步发展完善的理论，遵循一定的基础规律及长期实践总结出来的科学技巧。

二、翻译标准

科技英语的翻译标准应准确、流畅。

对于科技文献的翻译，准确是第一位的，在准确的基础上再求流畅。要特别注意逻辑和术语正确，结构严谨，表达简练。

三、科技英语的特点

1. 词汇

在科技英语中出现频率最高的不是专业词汇，而是一些功能词，如动词、介词、形容词等。

2. 语法

科技英语语法的突出特点如下。

(1) 被动句多（只强调过程本身，谁做的不重要）。

(2) 形容词后置定语多。

(3) 动词非谓语形式使用频率高。

(4) 祈使句和 it 句型多。

(5) 复杂长句和名词化从句（从句套从句、介词＋从句）多。

3. 翻译的特点

因为科技文献主要论述事理，其逻辑性强，结构严谨，术语繁多，所以译文必须满足概念清晰、条理分明、逻辑正确、数据无误、文字简练、通顺易懂等要求，尤其是翻译术语、定义、定理、公式、算式、图表、结论等时要注意准确、恰当。

附录 B　科技论文的英文摘要

摘要作为对科技论文正文的精炼概括，有利于读者在最短的时间内了解全文内容。随着国际检索系统的出现，摘要逐渐成为一种信息高度密集的相对独立文体，为人们在浩如烟海的文献中寻找所需要的信息提供了便利。

随着二次文献数据库的普及和全球科学技术界对科技信息日益增长的需求与重视，科技论文摘要受到的关注比论文本身多数十倍甚至数百倍。为此，一篇科技论文能否得到重视，能否把科研成果准确地传播出去，能否被更多重要的数据库收录，摘要的内容和质量起很大作用。

因为现在多数高等学校要求学生的毕业论文有英文摘要（含英文题目及关键词），所以有必要了解科技论文摘要的写作。

1. 定义与分类

abstract 即摘要、文摘。摘要和原文献在一起，置于正文前面；文摘是脱离于正文而独立存在的，如单独出版的文摘杂志中的文摘及情报系统存储和提供的文摘。

摘要是对文献的内容不加任何解释（interpretation）和评论（evaluation）的简要且准确的（concise and precise）表达（description）。

abstract、synopsis、summary 的区别如下。

（1）abstract：摘要、文摘，置于文前。

（2）synopsis：梗概，用于 movie、story、fiction 等。

（3）summary：概述，置于文末。

摘要一般分为两类：信息性摘要和说明性摘要。目前绝大部分科技期刊和论文都要求作者提供信息性摘要。

（1）信息性摘要（informative abstract）。

信息性摘要强调尽量多而完整地报道原文献中的具体内容，特别是研究目的、研究问题、研究方法和手段、主要论点和发现、得出的结论及建议、措施等。它包括文献的主旨和数据等，适用于学位论文（dissertation）、学术刊物论文（journal paper）、学术会议论文（conference paper）、展示论文（poster）等。

（2）说明性摘要（descriptive abstract or indicative）。

说明性摘要提供主要内容（problem or issue），但不介绍具体内容（content），适用于讨论性文献等。

2. 作用及特点

摘要便于读者搜索、查阅、浏览文献。

摘要独立于正文，通常收录于相应学科的摘要检索类数据库或专刊，撰写好摘要对科技论文是否被数据库收录和他人引用至关重要。

摘要为读者提供关于文献内容的有用信息，即文献包含的主要概念和讨论的主要问题。读者可从摘要中获知作者的主要研究活动、研究方法和主要研究结果及结论。它可以帮助读者判断此文献对研究工作是否有用，是否有必要获取全文，为科研人员、科技情报人员及计算机检索提供方便。

摘要的特点如下：

（1）独立性。摘要包括使读者理解原文献的基本要素，可离开原文献而独立存在。

（2）概括性。摘要把一篇文献的精华部分以精炼的文字、极短的篇幅概括出来，成为浓缩的信息。

（3）客观性。摘要是对文献不加评论和解释的客观报道。

3. 基本内容及形式

摘要的基本内容包括研究目的、研究方法、结果、结论及建议等。重点是结果和结论，即应突出科技论文的创造性成果和新见解。

英文摘要的单词数一般为 100～250，而学术论文、长篇报告摘要的单词数一般为 500，都仅限于一页。

摘要以主题句开头，阐明原文主题，但注意第一句不得与标题（title）重复，以免检索系统收录后有关人员用计算机检索时出现差错。

摘要一般不分段。当学位论文较长时可分段。

摘要用完整的句子，而不用电报式文字、短语，注意前后应连贯。

摘要尽量用主动语态，有时也可用被动语态。

摘要常用第三人称，若必须则可用第一人称；尽量用正式的书面用语，而不用口语化的词汇；尽量用简单的词汇，而不用复杂、生僻的词汇。

摘要应避免使用不常用的术语、首字母缩写词及缩略语等。

摘要最好最后撰写，以简洁地表述重要内容。

4. 时态问题

图表介绍、公式说明、实验结果、方法描述及客观真理等用一般现在时。

实验过程、过去做的研究工作、特殊结论及推论等用一般过去时。

叙述从某一时间开始，对现在有直接关系及影响的用现在完成时。

今后的研究及打算、预期的结果、数学公式推演结果等一般用将来时。

5. 精炼英文摘要的方法

（1）取消不必要的字句。

（2）尽量简化一些冗长的措辞，示例见附表 B-1。

附表 B-1　冗长的措辞示例

不用	用
at a temperature of 250℃ to 300℃	at 250℃—300℃
at a high pressure of 2000 Pa	at 2000 Pa
at a high temperature of 1500℃	at 1500℃
discussed and studied in details	discussed

（3）删除多余的文字。

错误：Type B, Type A, and Type C viral hepatitis accounts for 65.2%, 20.6%, and 13.2%, respectively.

正确：Type A, B, and C viral hepatitis accounts for 65.2%, 20.6%, and 13.2%, respectively.

后两个 Type 是不必要的重复，应删除。

（4）动词短语改单词。

① 用正规的动词而不用动词短语。

在科技英语中，常用规范的书面语动词代替口语中的动词短语。因为单个动词语义明确，而动词短语有时多义，甚至容易产生歧义，示例见附表 B-2。

附表 B-2 动词短语改单词示例

Phrasal Verb	Single Verb
take in	absorb
put together	aggregate
fall down	precipitate
get rid of	eliminate
join together	combine
break up	decompose
spread out	diffuse
come out	emerge
make sure	ensure
get away	escape
turn … into vapor	evaporate
take out	extract
grow longer	lengthen
turn … into liquid	liquefy
make wet	moisten
make neutral	neutralize
look at	observe
pour out over the top	overflow
move backwards and forwards in a straight line	reciprocate
throw back	reflect
set free	release
go up	rise
put back	replace
grow shorter	shorten
become a solid	solidify
stay alive	survive

续表

Phrasal Verb	Single Verb
pass on	transmit
make…weak	weaken

② 如果能用一个单词来表达意思的就不用词组表达。

应多用单个词而不用烦琐的词组，以使论文简洁明了，简化词组的首选用法见附表 B-3。

附表 B-3 简化词组的首选用法

Verbose Phrases	Preferred Usage
a considerable amount of	much
a considerable number of	many
a decreased amount of	less
a decreased number of	fewer
a majority of	most
a number of	many
a small number of	a few
absolutely essential	essential
accounted for by the fact	because
adjacent to	near
along the lines of	like
an adequate amount of	enough
an example of this is the fact that	for example
an order of magnitude faster	10 times faster
be apprise of	inform
are of the same opinion	agree
as a consequence of	because
as a mater of fact	in fact(or leave out)
as a result of	because
as is the case	as happens
as of this date	today
as to	about(or leave out)
at the rapid rate	rapidly
at an early date	previously
at no time	never
at some future time	later
at the conclusion of	after
at the present time	now

Verbose Phrases	Preferred Usage
at this point time	now
based on the fact that	because
because of fact that	because
by means of	by, with
causal fact	cause
be cognizant of	realize
completely full	full
consensus opinion	consensus
considerable amount of	much
be contingent upon	depend on
definitely proved	proved
despite the fact that	although
due to the fact that	because
during the course of	during
during the time that	during
enclosed herewith	enclosed
end result	result
entirely eliminate	eliminate
fatal outcome	death
fewer in number	fewer
first of all	first
for the purpose of	for
for the reason that	since, because
future plans	plans
give an account of	describe
give rise to	cause
has been engaged in a study	has studied
has the capability of	can
have the appearance of	look like
having regard to	about
important essential	essential
in a number of case	some
in a position to	can, may
in a satisfactory manner	satisfactorily
in a situation in which	when
in a very real sense	in a sense(or leave out)

续表

Verbose Phrases	Preferred Usage
in almost all instances	nearly always
in case	if
in close proximately to	close, near
in connection with	about, concerning
in light of the fact that	because
in many case	often
in my opinion	I think
assumption that	if
in only small a number of cases	rarely
in order to	to
in relation to	toward, to
in respect to	about
in the absence of	without
in the event that	if
in the not-too-distance future	soon
in the possession of	has, have
in this day and age	today
in view of the fact that	because, since
in as much as	for, as
incline to the view	think
is defined as	is
is desirous of	wants
it has been reported by Smith	Smith reported
it is apparent that	apparently
it is believed that	I think
itis clear that	clearly
it is clear that much additional work will be required before a complete understanding	I don't understand it
it is crucial that	must
it is doubtful that	possibly
it is evident that a produced b	a produced b
it is generally believed	many think
it is my understanding that	I understand that
it is of interest to note that	leave out
it is often the case that	often
it is suggested that	I think

Verbose Phrases	Preferred Usage
it is worth pointing out in this context that	note that
it may be that	I think
it may, however, be noted that	but
it should be noted that	note that (or leave out)
It was obeserved in the course of the experiments that	we observed
join together	join
lacked ability to	couldn't
large in size	large
let me make one thing perfectly clear	a snow job is coming
majority of	most
make reference to	refer to
meet with	meet
militate against	prohibit
more often than not	usually
needless to say	leave out
new initiatives	initiatives
no later than	by
of great theoretical and practical importance	useful
of long standing	old
of the opinion that	think that
on a daily basis	daily
on account of	because
on behalf of	for
on no occasion	never
on the basis of	by
on the grounds that	since, because
on the part of	by, among, for
on those occasions in which	when
our attention has been called to the fact that	we belatedly discovered
owing to the fact that	since, because
place a major emphasis on	stress
pooled together	pooled
presents a picture similar to	resembles
previous to	before
prior to	before

续表

Verbose Phrases	Preferred Usage
protein determinations were performed	proteins were determined
quite a large quantity of	much
quite unique	unique
rather interesting	interesting
red in color	red
refered to as	called
regardless of the fact	even though
relative to	about
resultant effect	result
root cause	cause
serious crisis	crisis
should it prove the case that	if
smaller in size	smaller
so as to	to
subject matter	subject
subsequent to	after
take into consideration	consider
the great majority of	most
the opinion is advanced that	I think
the predominate number of	most
the question is to whether	whether
the reason is because	because
the vast majority of	most
there is reason to believe	I think
they are the investigators who	they
this result would seem to indicate	this result indicates
through the use of	by, with
to the fullest possible extent	fully
unanimity of opinion	agreement
until such time	until
very unique	unique
was of opinion that	believed
ways and means	ways, means(not both)
we have insufficient knowledge	we don't know
we wish to thank	we thank
what is the explanation of	why

续表

Verbose Phrases	Preferred Usage
with a view to	to
with reference to	about(or leave out)
with regard to	concernssing, about(or leave out)
with respect to	about
with the possible exception	except
with the result that	so that
with in the realm of possibility	possible

6. 英文摘要的常用句型

（1）本文介绍了（描述了，阐述了）……

This paper describes…

This paper gives an account of…

In this paper, …is introduced

This paper treats of…

This paper is concerned with…

e. g. This paper is concerned with the derivation of optimum data of force.

（2）本文提出了……

This paper proposes/develops/extends/provides/presents…

e. g. This paper proposes/develops a new approach for the analysis of tolerance.

（3）本文展示了……

This paper shows…

e. g. This paper shows results on multiple time-scale system.

（4）本文分析了……

This paper analyses…

（5）本文研究了……

This paper considers/studies/deals with…

e. g. This paper considers the design of controllers for flexible systems.

（6）本文深入研究了……

This paper presents a thorough study of…

e. g. This paper presents a thorough study of the theory of temperature measurement.

（7）本文研究和分析了……

This paper studies and analyses…

（8）本文旨在研究……

The purpose of this paper is to explore/study…

e. g. The purpose of this paper is to explore the contact length between grinding wheel and workpiece in grinding.

（9）本文讨论了……

This paper treats of/discusses…

e. g. This paper treats of an important problem in date-base management systems.

This paper discusses the relation ship between the sampling period and the stability of sampled-data.

（10）本文论述了……

This paper addresses…

e. g. This paper addresses problems in linear quadratic optimal control of EHT transmission line.

（11）本文报告了……

This paper reports…

（12）本文论证了……

This paper establishes…

（13）本文总结了……

This paper reviews…

（14）本文给出了……

This paper presents/gives out…

（15）本文调查了……

This paper investigates/makes investigation on…

（16）本文指出了……

This paper points out/indicates that…

（17）本文的结论是……

This paper concludes that…

（18）本文将……和……作比较

This paper compares…with…/makes a comparision between…and…

e. g. This paper compares milling force with grinding force.

（19）本项研究的目的是……

The object of this study is…

（20）本文给出了一个……新方法

This paper gives a new approach…

（21）作者建议……（作者提出了……）

It is suggested that…（The author puts forward…）

（22）本文由……组成

This paper consists of/is composed of/has…

（23）闻起来……

It smells…

尝起来……

It tastes…

感觉起来……

It feels…

看上去……

It looks…

(24) 这种（现象）称为……

It is known as…

叫作……

It is called…

被看成……

It is referred to as

(25) 据说，据报道，据推算……

It is said/reported/estimated/calculated…

(26) ……是重要的/必要的/可能的/自然的/不可避免的

It is important/necessary/possible/natural/inevitable ＋ that（从句）或 to（不定式）…

(27)（我们）认为……是重要的/可能的/必要的/自然的/理所当然的

We think/consider it important/possible/necessary/natural that（从句）……或 to（不定式）……/We take it for granted that…

e. g. We think it necessary to design and build a set of machine for this kind of workpiece.

(28) 因为……（所以）……

because of…/owing to…/on account of…

e. g. Alloys are important because of (owing to) their usefulness in industry.

(29) 不管（无论怎样）……

No matter how/what/when/where/whether…

However/whatever/whenever/wherever…

e. g. No matter what this material is composed of, it is sure to deform if subjected to such a high pressure.

e. g. No matter how elastic a material is, it will never return its original shape or size once it is stretched beyond its limit of elasticity.

(30) 如果（假定，设……就……）

If…（should…had…）/Let…/Given…/With…/Without…

e. g. If there were no friction/Without friction, a machine would never stop operating once it is started.

It is impossible for this generator to wrong. If it should go out of order (should it go out of order), we could use the emergency one.

(31) 不但……而且……

not only… but also…

(32) 既不……又不……

Neither…nor…

不是……就是……

either…or…

e. g. This is an alloy which will neither expand nor contract when heated or cooled.

（33）……就……（一……就……）

as soon as

no sooner than…

（34）正是……

It is that…

7. 英文科技论文标题的特点和书写要求

英文科技论文的标题应简短扼要，不得超过 12 个单词。

英文科技论文的标题特点如下。

（1）词组多且多为名词性词组，一般为中心词＋修饰语。一般不用句子，如 Thermal Energy at Deep Mines，Methods to Reduce SteelWear in Grinding Mills。

（2）用冒号突出主旨，如 Advanced Vocabulary Learning，the Problem of collocation Vocabulary：Learning to be Imprecise.

英文科技论文标题的书写要求如下。

（1）除冠词、连词及少于 5 个字母的介词外，其余所有实词的首字母均应大写；当介词、冠词、连词在题目开头或最后一个单词时，其首字母也应大写（但有些刊物规定，题目的所有字母都大写）。

（2）字数在 20 字以内。若需分行，则第二行的第一个单词的首字母大写。

（3）不用或少用标点符号，除破折号和冒号外。

（4）每个单词都要有助于编制题录、索引和关键字等。

（5）力求简、明、短，不用完整的句子，多用名词性词组。

英文科技论文标题的常用词组如下。

（1）……的初步研究

Preliminary Study of…

（2）……的研究

A Study on…/A Study of the…/Studies on…/The Study of…

（3）……的实验研究

An Experimental Study on…/Experimental Study for…/Investigation on…

（4）……的理论和实验研究

A Theoretical and Experimental Study on…

（5）……的研制及其应用

The Research and Application of the…

（6）……的探讨

Exploration on …/Approach on …/Discussion on …/A study of …/Study on …/Investigation on…

（7）……的新进展

Recent Advances in…/Recent Progress in…

(8) ……的研究报告

Research Report on…

(9) ……的方法

Methods of…

……的计算方法

A Computational Method for…

(10) ……的分析

The Analysis of…

(11) ……的测量与分析

The Measurement and Analysis of…

(12) ……方法的改进

An Improvements on the Method of

(13) ……的调查

Investigation on…/Survey on…

(14) ……的设计

Design of…

(15) ……的合理化设计

The Rationalized Design of…

(16) ……的优化设计

Optimization Design of…/Optimum Design of…

(17) ……的设计准则

A Design Criteria for…

(18) ……的设计与研制

Design and Development for…

(19) ……的实用方案

Practical Scheme of…

(20) ……的试验

Experiments on…/Test of…

8. 摘要实例

波形铣刀片的开发及其铣削力学模型

Waved-Edge Insert Development and Milling Force Model

摘　要：在铣削机理研究和实验的基础上，开发了一种新型三维槽型铣刀片——波形铣刀片，并在直线刃铣刀片（含大前角铣刀片）铣削力学模型的基础上，建立了波形刃铣刀铣削力模型，同时编制了计算机程序进行预测。

Abstract：The waved-edge insert is developed based on the investigation of the milling mechanism and experiments. The milling force model for the wave-edge insert（including the great rake insert）provides the theoretical basis for developing, designing, and optimizing insert grooves through computer forecasting.

附录 C　参 考 译 文

第 1 部分

第 1 课　机械设计导论

机械设计涉及制订机械产品具体设计方案及解决具体机械工程问题,其目的是使设计的产品满足设计要求。它涉及的学科有材料学、力学、热学、流体学、控制学和电子学等。

机械设计可能简单,也可能复杂;可能容易,也可能困难;可能要求精确,也可能不要求精确;要解决的可能是一些平常琐碎的问题,也可能是非常重大的问题。好的设计源于有条理的和令人感兴趣的想法,并能提供一些成果或效用;一个好的设计产品应该是实用、高效且可靠的。好的产品与时常出问题和经常修理的拙劣产品相比,更经济实用。

进行机械设计工作的人通常称为工程师。他们必须首先认真定义问题,并用工程学的方法确保提出的任何一种方案都能解决问题。对于设计者来说,一开始就能准确制订出令人满意的设计方案,并能加以区别,以便选择一个最佳设计方案,这一点很重要。工程师必须具有创造性的想象力,熟悉工程知识、生产技术、工具、机器和材料,以设计、制造新产品,或者改进现有的产品。

在现代工业化社会,一个国家的财富和生活水平与其设计和制造工程产品的能力紧密相关,可以说机械设计和制造业的进步能显著促进一个国家工业化整体水平的提高。我国在全球制造业大舞台上正扮演着越来越重要的角色,为了促进工业化进程,需要越来越多具有广泛知识和专门技术的高技能设计工程师。

1. 机械零件

机器的主要部分是机械系统,机械系统可以分解成机构,机构可以进一步分解成机械零件。因此,机械零件是构成机器的基本单元。总体来讲,机械零件可以分为通用零件和专用零件两类。螺栓、齿轮和链条是典型的通用零件,它们广泛应用在各种工业部门的不同机器上;涡轮机的叶片、曲轴和飞机的螺旋桨是专用零件,它们具有一些特殊用途。

2. 机械设计步骤

产品的设计需要不断探索和发展。许多方案必须经过研究、试验、完善后才能决定是否采用。虽然每个工程学问题的内容都是独特的,但是工程师可以按照类似的步骤解决问题。机械设计步骤如图 1.1 所示。

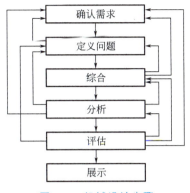

图1.1 机械设计步骤

(1) 确认需求。

设计是从设计者有了某种需求并决定对此进行研究开始的。这种需求常常不是很明显，通常是一种特殊的不利因素或一系列随机情况引起的，它们几乎同时发生，需求的确认通常是对尚不明确的问题的阐述。

(2) 定义问题。

定义问题对全面认识和理解问题非常必要，然后可以用更加合理和可行的方式重新阐述问题。问题的定义必须包括设计任务的所有技术条件，其中重要的要素有速度、进给量、温度限制、最大范围，以及变量、尺寸和质量的预期变化范围。

(3) 综合。

综合就是对各种可供选择的设计方法的进一步研究，此时往往不考虑这些方法的设计特点和设计质量。该步骤有时也称构思和发明阶段，在这个阶段中要产生尽可能多的有创意的想法。综合阶段包括材料的说明、几何特征的增加及总体设计尺寸的细节介绍。

(4) 分析。

分析是一种把问题分解成若干部分来确定或描述事物性质的方法。在这个过程中，需要分析设计的性质和原理，以便在修改的设计目标和原始设计目标之间确定一个更加合适的方案。

(5) 评估。

评估是对成功设计的最后验证，通常包括样机的实验室试验。在这个阶段，我们希望发现这个设计是否真的能够满足需求。

上述描述可能会给我们一个错误印象——这个过程可能像列出来的那样以线性方式来完成。事实恰恰相反，在整个过程中都需要反复进行，可能会随机从任一步骤返回到以前的某个步骤。

(6) 展示。

与其他人交流设计成果是设计过程至关重要的最后阶段。基本上有三种交流方式，分别是书面表述、口头陈述和图解。一名成功的工程师除应掌握技术外，还应精通这三种交流方式。有能力的工程师不应害怕在提出自己的方案时遭遇失败。实际上，失败乃成功之母。

3. 机械设计内容

"机械设计"是机械工程教学中的一门重要技术基础课程。它的宗旨是提供在机械设备和系统中的机械零件的设计所需的概念、规则、数据和决定性分析技术,并培养工程类专业学生的机械设计能力,这对机械制造业至关重要,也是制造好产品的关键。

机械设计包括以下内容。

(1) 设计过程、计算公式及相应的安全设计指标。

(2) 材料特性、静态和动态载荷分析,包括梁、振动和冲击载荷。

(3) 应力的基本规律和失效分析。

(4) 静态失效理论和静态载荷下断裂力学分析。

(5) 疲劳失效理论和强调在压力条件下接近高循环疲劳设计,通常用在旋转机械设计中。

(6) 机械磨损机理、表面接触应力和表面疲劳现象。

(7) 使用疲劳分析技术核校轴的设计。

(8) 润滑油膜与滚动轴承的理论和应用。

(9) 直齿圆柱齿轮的动力学原理、设计方法和应力分析,斜齿轮、锥齿轮和蜗轮等问题的简单介绍。

(10) 弹簧设计,包括螺旋线压缩弹簧、拉伸弹簧和扭转弹簧的设计。

(11) 螺钉、螺杆等紧固件设计,包括传动螺杆和预加载紧固件的设计。

(12) 盘式离合器和鼓式离合器及制动器的设计和技术说明。

第 2 课 基 本 概 念

在开始研究工程力学之前,理解某些基本概念和原理的意义是非常重要的。

1. 基本量

以下四个基本量在工程力学中常用。

(1) 长度。长度用于定位空间中点的位置,从而描述物理系统的大小。只要定义了一个标准的长度单位,就可以用该长度单位的倍数来定义距离和物体的几何属性。

(2) 时间。时间被认为是事件发生的一个连续系列。虽然时间与静力学的原理无关,但它在动力学的研究中起重要作用。

(3) 质量。质量是物质数量的一种度量,用来研究一个物体与另一个物体的相互作用。质量表现为两个物体之间的万有引力,并提供了物体阻碍速度变化的一种测量方法。

(4) 力。一般来说,力被认为是一个物体对另一个物体施加的"推"或"拉"的作用。这种相互作用可以发生在物体直接接触时,也可以在物体物理分离时隔着一段距离而发生,如引力、电场力和磁力。在任何情况下,一个力由它的大小、方向和作用点完全描述。

2. 理想化

模型或理想化在力学中用于简化理论的应用。下面介绍三个重要的理想化概念。

(1) 质点。质点具有质量,但其大小可以忽略。例如,由于地球的大小相对于其运行轨道的大小是微不足道的，因此，研究地球轨道运动时可以将它建模为一个质点。当一个物体被理想化为质点时，力学原理会得以简化，因为在问题分析中不会考虑物体的几何形状，如图2.1所示。

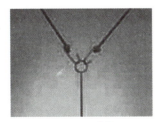

图2.1　质点

（2）刚体。刚体被认为是大量质点的组合，其中所有质点在施加载荷前后彼此之间都保持固定不变的距离。这是一个很重要的模型，因为当施加载荷时，物体的形状不会发生改变，所以不需要考虑制造物体的材料。大多数情况下，在结构、机器、机构等类似的体系中实际发生的变形量较小，此时刚体假设适用于问题的分析。

（3）集中力。集中力表示假定载荷作用于物体上的某一点。可以用一个集中力表示载荷，前提是载荷所施加的面积相对于物体的整体来说非常小，如车轮与路面之间的接触力，如图2.2所示。

图2.2　集中力

3．牛顿运动定律

工程力学是在牛顿三大运动定律的基础上阐述的，其正确性基于实验观察。这些定律适用于从惯性（非加速）参照系中测量的质点运动。

（1）牛顿第一定律。最初处于静止状态或匀速直线运动状态的质点，如果没有受到非平衡力的作用，则倾向于保持这种状态，如图2.3所示。

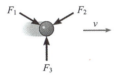

图2.3　平衡

（2）牛顿第二定律。受非平衡力 F 作用的质点会受到与力方向相同的加速度 a，其大小与力成正比，如图 2.4 所示。如果 F 作用于质量为 m 的质点，则牛顿第二定律可以用数学式表示为 $F=ma$。

图 2.4　加速运动

（3）牛顿第三定律。两个质点之间的作用力与反作用力是大小相等、方向相反且共线的，如图 2.5 所示。

图 2.5　作用与反作用

4. 牛顿万有引力定律

牛顿在假设了三大定律后，又假设了一个表述任意两个质点之间万有引力的定律，用数学式表示为

$$F=G\frac{m_1 m_2}{r^2} \qquad 式(2-1)$$

式中，F 为两个质点之间的万有引力；

　　G 为万有引力常数，根据实验数据，$G=6.673\times 10^{-11}\,\mathrm{m^3/(kg\cdot s^2)}$；

　　m_1 和 m_2 分别为两个质点的质量；

　　r 为两个质点之间的距离。

5. 重量

根据式(2-1)，任意两个质点之间都有一个相互吸引的力（引力）。然而，对于位于地球表面或靠近地球表面的质点，唯一具有相当大数量的引力是地球和质点之间的引力。因此，这种被称为重量的力是力学研究中唯一需要考虑的引力。

从式(2-1)可以推导出一个近似表达式，以便求出具有质量 $m_1=m$ 的质点的重量 W。假设地球是一个具有质量 $m_2=M_e$、密度均匀的非旋转球体，如果 r 是地球中心到质点的距离，则有

$$F=G\frac{mM_e}{r^2}$$

令 $g=GM_e/r^2$，得

$$W=mg$$

与 $F=ma$ 相比，可以看出 g 是由重力引起的加速度。因为它取决于 r，所以一个质点的重量并不是固定不变的，它由测量的地点决定（图 2.6）。对于大多数工程计算，g 是在纬度 45°的海平面处取得的，这个位置称为标准位置。

图 2.6 重量

此宇航员的重量减轻了,这是因为其远离地球的引力场

第 3 课 表 面 张 力

人们可以观察到,一滴血在水平玻璃上形成隆起;一滴水银形成一个接近球体的形状,可以像钢球一样滚过光滑的表面;悬挂在树枝上的雨水或露水、喷射到发动机的液体燃料、漏水的水龙头滴出的水滴、释放到空气中的肥皂泡、树叶上的水珠都近似球体[图 3.1(a)]。水黾具有非凡的"不湿之腿",故而能够毫不费力地在水面站立和行走[图 3.1(b)]。

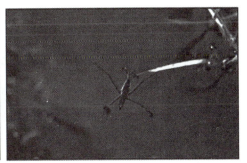

(a) 树叶上的水珠　　　　　　　　(b) 水面上的水黾

图 3.1 表面张力的作用效果

在这些观察中,液滴就像充满液体的小气球,液体的表面就像在张力下拉伸的弹性薄膜。液体分子之间的吸引会产生这种张力,其方向与液体表面平行。这个力在单位长度上的大小称为表面张力,用 σ_s 表示,单位为 N/m。这种效应称为单位面积上的表面能,以等效单位 $(N \cdot m)/m^2$ 或 J/m^2 表示。在这种情况下,σ_s 表示完成拉伸以增加一个单位的液体表面积所做的功。

为了可视化表面张力的产生过程,可以通过考察一个在液体表面而另一个在液体内部的两个液体分子实现,图 3.2 展示了它们的微观情景。因为液体上面的气体分子所施加的吸引力通常都很小,作用在液体表面分子上的吸引力是不对称的,所以在液体表面分子上存在一个吸引力的合力，它倾向于将液体表面分子拉向液体内部。这种力试图压缩液体表面以下分子，吸引力与排斥力平衡。这就是液滴倾向于形成球体的原因。球体具有给定体积下的最小表面积。

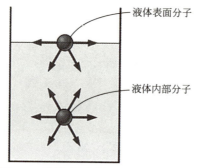

图 3.2　液体表面分子和液体内部分子

人们有时会观察到一些有趣的现象，如：有些昆虫可以趴在水面上，甚至可以在水上行走；细小的钢针可以漂浮在水面上；等等。这些现象都是通过表面张力平衡物体的质量实现的。

为了更好地理解表面张力效应，可以考虑用具有可移动侧边的 U 形钢丝框架拉动液体薄膜（如肥皂泡）的方法，如图 3.3 所示。通常情况下，液体薄膜倾向于将可移动钢丝拉向内侧，以使其表面积最小化。因此需要在可移动钢丝上施加一个方向相反的力 F，以平衡这种拉动效果。液体薄膜两侧表面都暴露在空气中，因此在这种情况下，表面张力作用的长度为 $2b$。由可移动钢丝上的力平衡可以得出 $F = 2b\sigma_s$，则表面张力

$$\sigma_s = \frac{F}{2b}$$

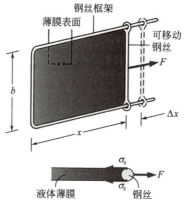

图 3.3　用 U 形钢丝框架拉动液体薄膜

注意到，对于 $b = 0.5\text{m}$，这时测量的力 F（单位为 N）等效于表面张力（单位为 N/m）。具有足够精度的这种仪器可以用来测量各种液体的表面张力。

在 U 形钢丝框架中，通过拉动可移动钢丝来拉动液体薄膜以增加其表面积。当可移动钢丝拉动距离为 Δx 时，表面积增加 $\Delta A = 2b\Delta x$，在此拉动过程中所做的功

$$W = F\Delta x = 2b\sigma_s \Delta x = \sigma_s \Delta A$$

假设在拉动距离很小时内力保持不变,在拉动过程中薄膜的表面能在数量上增加了 $\sigma_s \Delta A$,这类似于橡皮筋在被进一步拉伸后具有更多的(弹性)势能。对于液体薄膜的情况,所做的功用来克服其他分子的吸引力将液体分子从内部拉到表面。因此,表面张力也可以定义为液体增加单位表面积所做的功。

表面张力因物质不同而变化很大,并且对特定的物质来说,其随温度变化也很大。例如,在20℃下,被大气包围的水的表面张力为0.073N/m,而水银为0.440N/m。由于水银的表面张力足够大,因此水银可以形成接近球体的形状,像固体球一样在光滑的表面上滚动。一般来说,液体的表面张力随温度的升高而降低,在临界点处变为零(在临界点以上没有明显的液-气界面)。压力对表面张力的影响通常可以忽略不计。

由于物质的表面张力可以被杂质大大改变,因此可以添加某些被称为表面活性剂的化学物质到液体中,以降低其表面张力。例如,肥皂和洗涤剂降低了水的表面张力,使其能够渗入纤维之间的微小缝隙,以实现更佳的洗涤效果。但这也意味着,依赖表面张力而工作的设备,可能会因存在不良工艺导致的杂质而不能正常工作。

第4课 机械零件(I)

1. 齿轮

齿轮副由相互直接接触的齿轮组成,通常成对使用,在轮齿连续啮合的作用下,齿轮将运动和力从一根转轴传递到另一根转轴上,或者将运动和力从一根转轴传递到滑块(齿条)上。

齿轮的接触面必须在同一方向上对齐,从而使传动是正向的,也就是传递载荷不必依靠表面的摩擦作用。垂直于齿轮表面的公法线不能经过主动轮或从动轮的轴线。对于直接接触的实体(如齿轮),其摆线和渐开线轮廓均提供了正向的驱动及均匀的转速比,即产生共轭作用。

在一对齿轮副中较小的齿轮称为小齿轮,较大的齿轮称为大齿轮。当小齿轮安装在传动轴上时,这对齿轮用于减速;反之,当大齿轮安装在传动轴上时,这对齿轮用于加速。齿轮一般用于减速的情形要多于加速的情形。

如果齿轮有 N 个齿,并以 n r/min 的速度旋转,则乘积 $N \cdot n$ 表示每分钟旋转的齿数。如果每个齿都能在啮合区中与另一个齿轮的齿啮合,那么这个乘积对一对啮合齿轮的两个齿轮来说一定是相等的。

对于不同类型的共轭齿轮,齿轮比和速度比都可以通过大齿轮和小齿轮上的齿数比获得。如果一个大齿轮的齿数为100,小齿轮的齿数为20,齿数比为100∶20=5。无论大齿轮的转速为多少,小齿轮的转速都是大齿轮转速的5倍。大、小齿轮的接触点称为节点,由于节点位于中心线上,因此是齿形轮廓线上唯一做纯滚动的接触点。位于既不平行又不相交(交叉)的传动轴上的齿轮有节圆,但纯滚动节圆的概念对这种齿轮来说没有意义。

齿轮的种类一般是根据齿轮轴的排列方式划分的,这就意味着轴的排列布置形式确定后,齿轮的类型就大致定下来了。如果齿轮的速度变化及类型一定,那么轴的排列布置形式基本就定下来了。

轮廓为平行于其轮轴直线的齿轮称为直齿轮，直齿轮只用于轮轴平行的齿轮间的传动。假设一个渐开线直齿小齿轮是用橡胶制成的，并且能够均匀旋转，那么其两端会绕轴线做相对转动，开始是直的且平行于轴线的小齿轮上的齿就变成了螺旋形，即小齿轮变成了螺旋齿轮。

为了使交叉轴斜齿轮实现线接触并提高其承载能力，大齿轮可能被制作成部分弯曲的形状以围绕小齿轮，类似于螺母套在螺杆上，形成圆柱形的蜗轮蜗杆。蜗杆有时也制作成沙漏状，而非圆柱状，目的是使蜗杆部分接触蜗轮，进一步提高承载力。

蜗轮蜗杆是只用一对齿轮就可以提供较大转速比的装置，但由于沿齿面方向存在滑动现象，因此蜗轮蜗杆的传动总是比平行轴传动的齿轮效率低。

2. V形带

人造纤维和橡胶V形带广泛用于传递动力。V形带一般有两种：标准V形带和重型V形带。V形带用于传动中心距离较小的场合，可以制造成无缝的，避免连接设备的麻烦。

第一，V形带成本低，通过并排增加V形带可以增大传动功率。传动时，所有V形带都被拉伸相同的长度以保持每条V形带中的载荷均匀。当一条V形带断裂时，通常所有V形带都需要被调换。带传动中（皮带）的紧边可以在上面，也可以在下面，并以任意角度倾斜。由于V形带工作在一个较小的带轮中，因此采用一个带轮就可以达到较大的减速。

第二，带轮槽的倾斜角度通常为34°～38°。V形带在槽中的嵌入作用可以大大增大V形带的牵引力。

第三，带轮可以用铸铁、钢、锻压金属等材料制成。在带轮槽的底部需要留有足够的间隙，以保证V形带不接触带轮槽的底部。有时较大的带轮没有轮槽，靠带的内表面获得牵引力，从而节省在带轮上加工槽的成本。厂家供应出售的带轮可根据用户要求调节槽宽。带轮的节距是变化的，通过适当的调整可以实现所需的传动速度。

3. 链传动

链传动自行车最早出现在1874年，那时的链用于驱动早期自行车的后轮。如今，随着现代设计和制造方法的改进，链传动的应用越来越广泛，已经极大地提高了农业机械、钻探设备、矿业和建筑机械的效率。大约自1930年以来，链传动日益普遍，尤其在动力锯子、摩托车和自动扶梯等设备上应用广泛。

下面介绍三种动力传送链，分别为滚子链条、齿型链条或低噪声（无声）链条、滚珠链条。

（1）滚子链条。

滚子链条的基本组成部分有链板、销轴、套筒、滚子、两个或两个以上链轮，每个链轮上有类似齿轮形状的齿。滚子链条用销轴和滚柱等装配而成，用两个销轴插入两个链板中的孔，可以连接内、外链板，销轴紧紧地插入孔中，形成压紧连接。滚子由内、外链板和两个压紧的套筒组成，硬的钢质滚柱可以自由转动。装配后，销轴与套筒自由配合，随着链条在链轮上传动，销轴可相对于套筒轻微转动。

标准滚子链条可以是单排的,也可以是多排的。如果是多排的,那么两条或多条链条一起连接在同一个销轴上,这时要使滚子以一定的排列方式对齐,单个驱动的速度比一般控制在10:1,较合适的轴中心距离一般为30~35倍滚柱与滚柱的距离。链条速度一般不超过800m/min。当同时驱动多个平行轴时,滚子链条传动尤为合适。

(2) 齿型链条或低噪声(无声)链条。

齿型链条或低噪声(无声)链条基本上是齿轮齿条的组合件,每个齿条有两个齿,轴向连接形成一个封闭的、内面有齿的链条,链轮上有共轭齿。链条连接采用销子连接平钢板,平钢板上通常是倾角为60°的直齿,便于许多链条同时传递动力。相对于滚子链条来说,这种类型的链条噪声较低,能在较高速度下工作,同样的宽度能传递更大的载荷。汽车上通常采用这种低噪声链条。

(3) 滚珠链条。

滚珠链条用于连接平行或非平行轴之间的运动,连接的方式灵活,通常用在传动速度和动力较低的场合。链轮含有球形或锥形凹槽,小滚珠可以在凹槽内活动。用普通碳钢、不锈钢或实心塑料制成的滚珠安装在链条上。滚珠链条用于计算机、空调、电视调谐器、百叶窗帘等物品上。链轮可以用钢、压铸成形的锌或铝及单体浇注成形的尼龙等制成。

第5课 机械零件(Ⅱ)

1. 紧固件

紧固件可以将一个零件与另一个零件连接。因此,几乎在所有设计中都要用到紧固件。

紧固件有以下三个主要类型。

(1) 可拆式紧固件。可拆式紧固件容易拆卸,而且拆卸时不会对紧固件造成损伤,如用普通的螺栓、螺母连接的零件。

(2) 半永久式紧固件。虽然半永久式紧固件能被拆卸,但通常会对所用的紧固件造成损伤,如开口销。

(3) 永久式紧固件。永久式紧固件不能拆卸,如铆接和焊接的零件。

对于任何复杂产品来说,紧固件都非常重要。以汽车为例,它是由数千个零件装配而成的。一个紧固件的失效或松动可能会带来车门响等小麻烦,也可能造成车轮脱落等严重后果。因此,为一个特定用途选择紧固件时,应考虑上述各种可能性。螺母、螺栓和螺钉是连接材料最常用的零件。由于紧固件应用非常广泛,因此价格低、功效好是非常必要的。

当激振力大于摩擦力时,普通螺母会松动。在螺母与锁紧垫圈组件中,锁紧垫圈具有独立的锁紧特性以防止螺母松动。锁紧垫圈只有在螺栓与装配件的长度相对变化而松动时才起作用。这种螺栓长度的变化是多种因素引起的,如螺栓内部蠕变、弹性丧失、螺栓与被连接件的热膨胀差异或磨损。在上述静态情况下,膨胀的锁紧垫圈在轴向载荷的作用下固定螺母并确保紧固。当振动相对改变时,锁紧垫圈将不起作用。

铆钉属于永久式紧固件,依赖于结构的变形来固定。铆钉通常比螺纹连接件牢固,而且生产成本低。铆钉既可热铆又可冷铆，取决于铆钉材料的机械特性，如铝制铆钉是冷铆，因为冷却可以提高铝的强度；但大多数大型铆钉是热铆。

2. 轴

实际上，几乎所有机器中都装有轴。轴最常见的形状是圆柱形，其横截面可以是实心的，也可以是空心的（空心轴可以减轻质量）。

轴安装在轴承中，通过齿轮、皮带轮、凸轮和离合器等传递动力。通过这些零件传递的力可能会使轴发生弯曲变形，因此，轴应具有足够的刚度以防止支承轴承受力过大。总而言之，在两个支承轴承之间，轴在1foot（约12in，即0.30m）长度上的弯曲变形不应超过0.01in（约0.00025m）。

直径小于3in（0.08m）的轴可以采用含碳量约为0.4%的冷轧钢，直径等于3~5in（约0.08~0.13m）的轴可以采用冷轧钢或锻造毛坯。当直径大于5in（约0.13m）时采用锻造毛坯，然后切削加工到所要求的尺寸。塑料轴广泛应用于轻载场合。由于塑料是电的不良导体，因此在电气工程应用中用塑料轴比较安全。

在轴的设计中，轴与轴的连接方式是要重点考虑的。轴与轴之间的连接可由刚性或者弹性联轴器实现。

3. 轴承

轴承是一种经过特定设计并用于支承机器上运动部件的零件。轴承通常作为支承传递运动的转动轴。由于轴承和配合表面间存在相对运动，因此会产生摩擦。在许多场合（如设计皮带轮、制动器和离合器）下，摩擦是有益的。然而，设计轴承时，减小摩擦是要考虑的基本因素之一，因为摩擦会导致功率下降、发热和配合表面磨损等。

设计球轴承和滚子轴承时，机器设计人员应考虑五个方面：寿命与载荷的关系、刚度或在载荷作用下的变形、摩擦、磨损和噪声。对于中等载荷和转速，应根据额定负荷选择轴承，以保证轴承良好的工作性能。当载荷较大时，轴承零件会发生变形，尽管变形的程度远小于轴或其他与轴承相连的零部件，但对轴承零件仍然会有很大影响。在转速高的场合需要有专门的冷却装置，这可能会增大摩擦力。磨损主要是污染物的进入引起的，必须选用密封装置。

虽然球轴承和滚子轴承的设计由轴承制造厂负责，但机器设计人员必须对轴承的使用有正确的估计，不仅要考虑轴承的选择，还要考虑轴承的正确安装条件。

轴承套圈与轴或轴承座的配合非常重要，它们之间的配合不仅应该保证所需过盈量，而且应该保证轴承的内部间隙。不正确的过盈量会产生微动腐蚀，导致严重故障。内圈通常是通过紧靠在轴肩上进行轴向定位的。轴肩处的圆弧半径主要为了避免应力集中。在轴承内圈加工出的圆弧或者倒角可用来提供容纳轴肩处圆弧半径的空间。

滑动轴承最简单的形式是由合适的材料和加工尺寸制成的衬套。轴颈通常是旋转在轴承内的轴的一部分，如图5.1所示。

滑动轴承在工作中伴随滑动接触，为了减小滑动摩擦引起的问题，应使用与配合材料相对应的润滑油。选择润滑油和配合材料时，通常要考虑轴承压力、温度和滑动速度等因素。在滑动轴承中，润滑油的基本作用是阻止摩擦表面直接接触，因此，在不同载荷、速度和温度下，油膜维护是在滑动接触中考虑的首要问题。

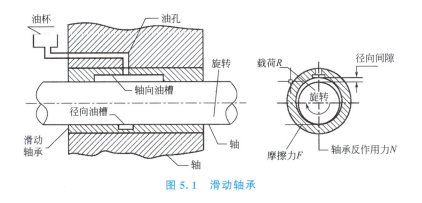

图 5.1 滑动轴承

第 6 课 机 构

机构是两个或两个以上构件通过活动连接以实现规定运动的构件组合，是机械的组成部分。两个有相对运动的构件间的活动连接称为运动副。其中，面接触的运动副称为低副，点或线接触的运动副称为高副。机构的运动特性主要取决于构件间的相对尺寸、运动副的性质及相互配置方式等。机构中用以支持运动构件的构件称为机架，可作为研究运动的参考坐标系。具有独立运动的构件称为原动件。机构中除机架和主动件外的被迫做强制运动的构件称为从动件。描述或确定机构运动所必需的独立参变量（坐标数）称为机构自由度。为使机构的构件获得确定的相对运动，必须使机构的原动件数等于机构自由度。

机构可以分为平面机构、球面机构和空间机构三类。这三类机构具有许多共同点，区别分类的标准在于连杆装置的运动特性。

在平面机构中，所有点在空间绘出的是平面曲线，所有曲线都在平行平面上，即所有点的轨迹都是与单一公共平面平行的平面曲线。根据这一特点，能够在单个图形或图像上以实际尺寸和形状绘出平面机构任意选择点的轨迹。这种机构的运动变换称为共面。目前最常用的机构是平面机构，如平面四连杆机构、平面盘形凸轮、从动件及曲柄滑块机构。图 6.1 所示为凸轮。

凸轮是驱动从动件实现某一具体运动的机械零件。恰当的凸轮结构设计可使机器零件获得预期的运动。凸轮广泛应用于几乎所有的机械设备中，包括内燃机、机床、压缩机和计算机等。一般来说，可按以下两种方式设计凸轮。

（1）设计凸轮轮廓，将预期的运动传给从动件。

（2）选择能满足从动件运动要求的凸轮轮廓。

旋转的凸轮可将柱面运动变为直线运动。凸轮的作用是把各种运动传输给机器的其他零件。

每个凸轮都必须根据特定需要设计和制造。虽然每个凸轮都看起来与其他凸轮有很大的不同，但是所有凸轮都以相似的方式工作。在不同情况下，当凸轮旋转或转动时，与凸轮接触的另一个零件称为从动件，做左右、上下或内外运动。从动件通常与机器其他零件连接以完成预期的运动。如果从动件与凸轮不接触，则不能工作。凸轮是按它们的基本形状来分类的。

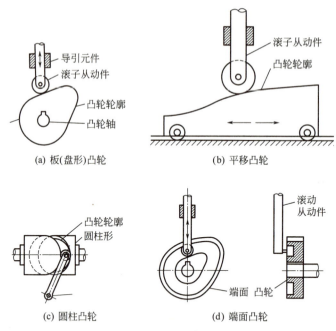

图 6.1 凸轮

凸轮按照形状分为四种：板（盘形）凸轮、平移凸轮、圆柱凸轮和端面凸轮。

仅采用低副的平面机构称为平面连杆机构，它可以只包括转动副和移动副；虽然理论上还可以包括平面副，但平面副不能加以约束，只相当于开式运动链。平面运动要求所有移动副的轴和转动副的轴都垂直于运动平面。

第 7 课　流体和液压系统

水力的由来已久，可以从人类开始利用能源时算起。世界上最容易利用的能源是两种可自由流动的流体——水和风。

水车是第一台液力装置，属于水力领域最早的发明物。15 世纪早期，水车图画就出现在拜占庭华丽宫殿的马赛克上。磨粉机由罗马人发明，而水磨机的历史更早，可以追溯到大约公元前 100 年。当时谷物的种植已经有 5000 多年的历史了，一些农场主厌烦了手工碾碎、研磨谷物的工作方式。但实际上，真正的水磨机发明家应是农场主的妻子，因为她们经常要干一些繁重的农活。

流体是可以流动的物体，也就是说，构成流体的粒子可以连续地改变它们之间的相对位置，而且随着流体层之间位置的改变（无论这个改变有多大），流体层之间的流动阻力将发生变化。

流体可以分为牛顿流体和非牛顿流体。在牛顿流体中，流体层之间作用的剪切力和角度变形量呈线性关系；在非牛顿流体中，流体层之间作用的剪切力和角度变形量呈非线性关系。

流体流动方式有很多种分类，如定常流或非定常流、有旋流或无旋流、可压缩流或不可压缩流、黏性流或无黏性流。

所有液压系统都遵守帕斯卡定律，该定律是以其发明者布莱士·帕斯卡（Blaise Pascal）的名字命名的。帕斯卡定律指出，在密封的容积（如圆柱筒或管子）内，压缩的液体在容积的各曲面上作用着大小相等的力。

在实际液压系统中，帕斯卡定律是解释系统内各种现象的依据。油泵驱使液体在液压系统中流动，油泵的进口连接油箱的液流源，作用在油箱液面上的气压驱动流体进入油泵。油泵工作时，流体在适当泵压的作用下，从油箱流动到出口。

油泵泵出的压缩液体由各种阀门控制。在大多数液压系统中，通常采用以下三种控制形式：液体压力控制、液体流速控制、液体流动方向控制。

下列几种情况通常优先使用液压驱动。

（1）动力传递距离大（链传动和皮带传动除外）的场合。

（2）需要低速、高转矩的场合。

（3）需要紧凑结构的场合。

（4）要求传动平稳、避免振动的场合。

（5）容易调节速度和方向的场合。

（6）需要输出无级调速的场合。

图 7.1 所示为液压电动机的速度控制。电力驱动的油泵驱使油液流动，将能量传递给液压电动机或液压油缸，从而将液压能转换成机械能。在阀门的控制下，压力油液产生线性或旋转的机械运动。油液流动的动能相对较低，因此有时采用静压传动。由于液压电动机和液压油缸之间几乎不存在构造差异，因此任何一个油泵都可以当作液压电动机使用。调节控制阀或变量泵可以控制在任一时间里的油液流量。

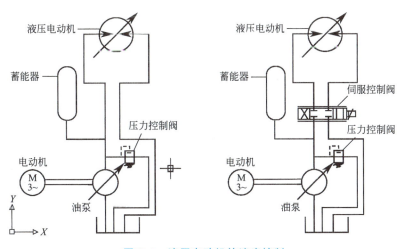

图 7.1 液压电动机的速度控制

虽然目前很多机床采用液压传动，但这并不是一种新的应用。实际上，现代独立油泵技术的发展已促使液压传动技术广泛应用在机床上。

液压驱动机床有许多优点。例如，采用液压驱动可以在较大的速度范围内向机床提供无级变速。此外，它能非常方便地改变驱动方向。在许多机床上，采用液压传动可以简化甚至完全不使用结构复杂的机械装置。

液压驱动的柔性和缓冲性能较好。除了能使机床运行平稳，液压驱动在机床上还可以改善工件表面的光洁度、可以在不损坏刀具的前提下增大刀具上的作用载荷、可以进行长时间的切削加工而无须再次刃磨刀具等。

第8课　工程图学

"图学"一词来源于希腊语"grapho"，其引申意为"绘图"或"图样"。图样是开发和交流技术思想的重要工具。工程图样可为零件提供准确、完整的描绘。除了物体的形状，工程图样还可以给出制造所需的全部资料，如尺寸、公差等。每位工程技术人员都必须学习投影知识，具有绘图和读图的能力、空间分析能力和空间想象能力。

作为工程界的共同语言，图样是用来指导生产和交流技术的。因此，必须对制图的各方面（如图样格式、图样画法、尺寸标注等）作出统一规定。

1. 三视图的形成

由于仅用一个投影无法确定空间点的位置，因此需要添加多个投影面。通常，三个相互垂直的投影面构成正投影面体系，分别是水平投影面、正投影面、侧投影面，记作 H、V、W。

三投影面将空间分为八个象限，如图8.1所示。依据GB/T 14692—2008《技术制图　投影法》，制图采用第一角画法，而美国和加拿大等国家采用第三角画法。在本课，我们采用第三角画法。

在第一角画法中，物体放在第一象限，观察者透过物体看投影面；在第三角画法中，物体放在第三象限，观察者透过投影面看物体，投影面被假想为透明的，从而形成视图。

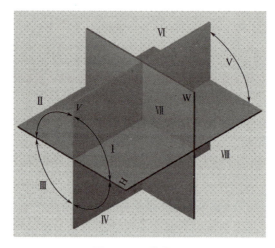

图8.1　八个象限

第一角画法和第三角画法如图8.2所示。

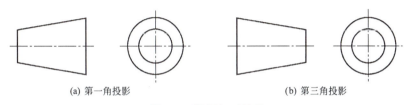

(a) 第一角投影　　　　　(b) 第三角投影

图 8.2　投影识别符号

2. 组合体

物体的投影规律：主视图和俯视图长对正、主视图和右视图高平齐、俯视图和右视图宽相等。

画组合体三视图的方法如下。

(1) 形体分析法：任何一个组合体都可以看作由一些简单的基本体组成。这些基本体都易于确定，可分为叠加式基本体和切割式基本体。

(2) 视图选择：主视图是表达组合体的一组视图中最主要的视图。确定主视图时，应选择最能反映组合体形状特征的方向作为主视图的投影方向。

(3) 画图步骤：定轴线、对称中心线和基线；用 H 铅笔画底稿；检查全图并加深图线。

阅读组合体视图时，将物体分解为若干单独的基本体；读图时，至少要把两个视图联系起来分析；明确视图中线和面的含义。

阅读组合体视图的方法有形体分析法和线面分析法。

(1) 形体分析法：将物体分解为若干基本体。

(2) 线面分析法：将物体分解为不同的线和面。

图 8.3 所示为组合体及其三视图。

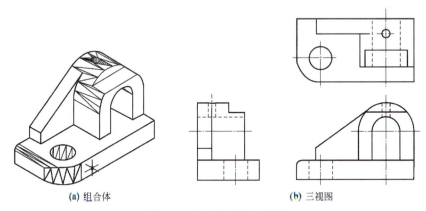

(a) 组合体　　　　　(b) 三视图

图 8.3　组合体及其三视图

3. 零件图

零件图是表示零件结构、大小及技术要求的图样。零件图描绘零件的形状、给出零件的尺寸，并提供生产该零件所需的全部资料。

零件图包含以下内容。

(1) 足够的视图，以完整地表达零件内、外结构的形状。

(2) 制造零件所需的全部尺寸。

（3）技术要求，包括形状公差、位置公差、表面粗糙度、材料规格和热处理要求等。

选择视图的方法如下。

（1）为了满足完整、清晰地表达物体形状的要求，在画图之前确定需要的视图和零件的最佳投影方案。

（2）选择主视图的原则有形状特征原则、功能位置原则、加工位置原则。

（3）选择其他视图的原则有用最少的视图表达最完整信息原则、避免使用虚线原则、避免细节的不必要重复原则。

4. 装配图

机器或部件的各零件按工作位置装配的图样称为装配图。

装配图包含以下内容。

（1）一组视图：表示在装配中各零件的位置关系和相互作用。

（2）少数尺寸：表示重要零件间的位置关系和产品的安装定位等所需的尺寸。

（3）技术要求：包括装配、检验和维修机器所需的信息。

（4）各零件的件号、明细栏和标题栏。

装配图的规定画法如下。

（1）一般画法。在装配图中，两接触表面或配合表面之间不画间隙，在非接触表面或非配合表面间应画出间隙；邻接零件的剖面线应从不同的方向或以不同的间隙画出；实心件（如转轴、心轴、杆件、手柄、销和键等）沿轴线被剖时不画剖面线，螺钉、螺栓、螺母及垫圈也不画剖面线。

（2）特殊画法。特殊画法包括沿结合面剖切或把某些零件拆开的画法、单独表示零件画法、使用假想线画法、夸大画法、简化画法。

第 9 课　CAD/CAM/CAPP 导论

在工业社会发展的进程中，人们已经开发出许多新技术和新产品。与以前出现的任何科学技术相比，对工程制造业冲击快、影响大的是计算机。如今，计算机已被广泛应用在工程产品的设计中。

计算机辅助设计（computer aided design，CAD）就是在产品设计过程中，利用计算机和图形软件对产品进行辅助设计，以提高产品设计效率的一种技术。在使用 CAD 的过程中，通常要用到一个交互式计算机图形系统，该系统称为 CAD 系统。CAD 系统是功能强大的工具，可用于产品及零部件的机械设计和几何建模。

使用 CAD 系统进行产品辅助设计的优点如下。

（1）提高产品的生产效率。

（2）提高产品的设计质量。

（3）统一产品的设计标准。

（4）建立产品的制造加工数据库。

（5）减少手工绘图（描图）出现的错误和图样间标准不统一等问题。

CAD 中的几何模型可以分为二维模型、三维模型和二维半模型。二维模型表示的是平面图形；三维模型一般能凸显机械零部件的形状特点（图 9.1）；二维半模型可以表示零部件单侧连续横截面的形状细节，它的主要优点在于无须建立完整的三维模型就可以表达零部件必要的形状特点。

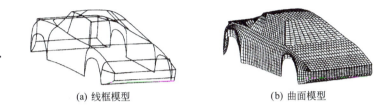

(a) 线框模型　　　　　　　　(b) 曲面模型

图 9.1　三维模型

在完成零部件详细的结构设计之后，需要对其进行功能性分析，这也是零部件设计的环节之一。功能性分析涉及零部件（或机构）的应力-应变分析、传热分析及动力学仿真分析等。在 CAD 系统中通常有一些机械零部件功能性分析的应用实例，如零部件的质量特性分析和有限元分析。质量特性分析包括计算几何体的体积、表面积、质量和重心等。在大多数 CAD 系统中可以进行有限元分析，主要分析内容包括对零部件的传热状况分析、应力-应变分析、动力学特性分析等。目前大多数 CAD 系统都能自动生成用于有限元分析的二维模型或三维模型。

计算机辅助制造（computer aided mapping，CAM）包括制造决策、生产过程和操作规划、程序设计、人工智能、自动化控制制造设备（如数控机床、数控加工中心、数控加工单元、数控柔性制造系统）及相应的技术（如数控单元技术、直接数控成组技术）等。

CAM 涵盖了成组技术、制造数据库技术、自动化和公差检测技术等。图 9.2 所示为 CAM 的总体框架。

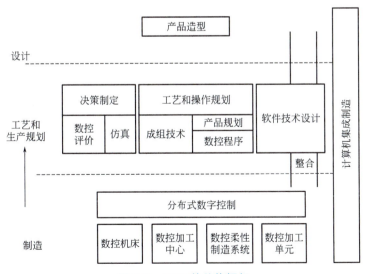

图 9.2　CAM 的总体框架

产品设计好之后，就要对其进行制造。如今，在机械产品制造中的许多环节里，计算机都扮演着重要的角色。例如，现代造船业中建造的结构都是由大型钢板上裁剪下来的钢板焊接成的，人们常使用计算机控制的火焰切割机；同时，人们利用计算机对船体部件的结构布局方式进行优化，以减少材料的浪费。

计算机辅助工艺设计（computer aided process planning，CAPP）是工艺员借助计算机制定产品工艺流程的过程，利用程度取决于制定流程的方法。低层次的CAPP是指工艺员利用计算机对手工完成的工艺数据进行存储和修改，同时为编制新的工艺流程提供数据。高层次的CAPP是指工艺员借助计算机自动完成一些形状简单工件的工艺流程，工艺员有时需要输入一些必要的数据，有时需要修改不符合特定生产要求的工艺流程。最高层次的CAPP也是CAPP的最终实现目标，它将工艺流程编制的理论、技术及经验编成程序输入计算机，计算机自动完成工艺流程的编制，无须工艺员参与。此外，它的数据库可直接与其他系统集成，如CAD系统和CAM系统，因此CAPP被认为是计算机集成制造系统中的一个重要组成部分。

第 10 课　工 程 公 差

物体的形状由轮廓决定。设计人员会给零件标注满足要求的公称尺寸。实际上，由于存在表面不规则及固有的表面粗糙度，不能完全按公称尺寸加工零件，因此必须允许尺寸有一些变动量以确保能够制造，但尺寸变化范围不能太大，以免装配性能变差。单个零件所允许的尺寸变化量范围称为公差。

1. 零件公差

为了确保零件的装配和互换性，有必要控制尺寸，对于影响间隙和过盈配合的关键尺寸应指定公差。指定公差的一种方法是在公称尺寸后标出允许偏差，如尺寸可以标注为 40.000 ± 0.003 mm，这意味着加工的尺寸必须为 39.997mm～40.003mm。误差可以在公称尺寸两侧变化的公差称为双向公差。单向公差的一个公差为零，如 $40^{+0.006}_{0}$。

当图中没有明确给出尺寸公差时，可以采用通用的尺寸公差。对于加工尺寸，通用公差可能是 ±0.5 mm。因此，被指定为 15.0mm 的尺寸范围可以是 14.5～15.5mm。其他通用公差适用于角度、钻孔、冲孔、铸件、锻件、焊缝及焊角等情形。

确定零件的公差时，可以参考以前的图样或普遍的工程惯例。公差通常以英式或ISO标准来定义范围。通用公差应用见表 10－1。

表 10－1　通用公差应用

分类	特性	ISO 标准码	装配	应用
自由转动	对大的温度变化、高转速或大的轴颈压力有益	H9/d9	有显著间隙	多轴承轴 液压缸内的活塞 活动杠杆 滚筒轴承
紧动配合	适用于在精密机械上转动及在中等速度和轴颈压力下精确定位	H8/f7	有间隙	机床主轴承 曲柄轴和连杆轴承 轴套 离合器套 导向块
滑动配合	适用于部件不用自由转动但必须精确移动、转动和定位的情况	H7/g6	推配合，没有显著间隙	推进齿轮和离合器 连杆轴承 指标活塞

续表

分类	特性	ISO 标准码	装配	应用
定位间隙配合	为固定件的定位提供密配合，但可以自由装配	H7/h6	手的压力、带润滑	齿轮 尾架套筒 调整环 活塞螺栓
定位过渡配合	适用于精确定位（位于间隙配合与过盈配合之间）	H7/k6	容易用锤子敲	滑轮 离合器 齿轮 飞轮
定位过渡配合	适用于更精确的定位	H7/n6	需要压力	电动机轴衔铁 轮子上带齿的轴环
定位过盈配合	适用于需要有定位精度和刚性的部件	H7/p6	需要压力	开式滑动轴承
中等驱动配合	适用于普通钢机件或小横截面上的紧缩配合	H7/s6	需要压力及温度差	离合器从动盘毂 滑块、轮子和连杆内的轴承轴瓦

2. 孔和轴的标准配合

在制造工程中常要求确定圆柱形零件的公差，如安装在相应的圆柱体零件或孔内的轴，或在其内部旋转的轴。配合的松紧取决于应用场合。例如，定位在轴上的齿轮需要紧配合，这里轴的直径实际上比齿轮轮毂的内径稍微大一些，以便传递所需转矩。另外，滑动轴承的直径必须大于轴的直径以使其旋转。假定从经济上讲不能把零件制造成精确的尺寸，则必须确定轴和孔尺寸上的变化量，但是变化的范围不应太大，以免装配时受损。与公差有关的术语如图10.1所示。

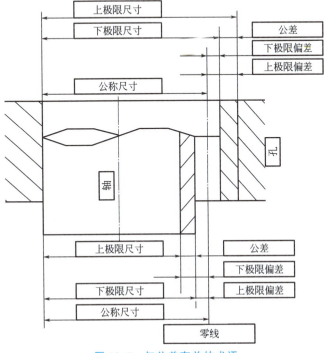

图 10.1 与公差有关的术语

尺寸：以专用单位表示的尺寸数值。

实际尺寸：通过测量得到的部件尺寸。

尺寸极限：零件允许的最大尺寸和最小尺寸。

上极限尺寸：两个尺寸极限中较大的。

下极限尺寸：两个尺寸极限中较小的。

公称尺寸：用以确定尺寸极限的参考尺寸。

偏差：尺寸与其对应的基本尺寸的差值。

上极限偏差：上极限尺寸与其对应的基本尺寸的差值。

下极限偏差：下极限尺寸与其对应的基本尺寸的差值。

公差：上极限尺寸与下极限尺寸的差值。

轴：用于常规设计一个零件所有外部特征的术语。

孔：用于常规设计一个零件所有内部特征的术语。

第 11 课　现代优化设计方法基础

　　如今，综合运用机械优化设计方法、有限元分析（finite element analysis，FEA）和 CAD 系统进行产品设计对设计过程产生了深远影响。这种综合运用的手段将工程师身上的设计重担交给计算机，降低了产品的设计成本。此外，正确运用优化设计中严谨的数学推理可以提高产品设计的可靠性。优化方法决定了产品设计过程中的精度问题，包括 CAD 系统几何建模的准确度、有限元分析中网格划分的准确度及分析处理器的计算精度等。这种方法能够在考虑机械、热等许多实际情况的影响下，对 CAD 系统构建的零部件、装配体的结构进行优化。

　　从优化设计理论的角度来说，CAD 文件和 FEA 文件之间无须任何格式转换就可以实现数据的无缝交换。只要这两个文件之间存在关联性，对 CAD 文件所作的所有修改就都会在相应的 FEA 文件中反映出来。例如，在使用有限元分析软件 ALGOR 对某零部件或装配体进行计算时，不需要建立其有限元模型就可以进行优化设计。用户只要挑选出零部件或装配体 CAD 模型中需要优化的几何尺寸，确定相应的设计准则（如最大应力、最高温度和最大频率），然后运行相应的分析计算过程，该软件通过计算、比较就可以完成 CAD 模型的结构优化，并且整个过程通常无须使用者参与。需要注意的是，CAD 文件与 FEA 文件的关联性使 FEA 模型有所更新，但约束和施加的载荷保持不变，因此需要对更新后的有限元模型进行计算，对整个过程不断重复迭代，直到最终计算结果满足设计要求为止。图 11.1 所示为零部件形状优化设计过程。

　　1. 机械结构的优化设计过程

　　在机械结构的优化设计过程中往往需要进行多步迭代计算，在整个计算过程中，零部件的几何外形得到不断优化。在每步迭代计算中都要进行一定的分析，以便得到与工程实际较相符的设计结果。优化设计一般需要很多步这样的迭代计算，每步计算都较费时。所以，在进行机械结构优化设计过程中使用优化设计软件的主要目的是自动运行上述迭代计算，减少工程师的工作负担。乍看上去，优化设计技术是一种能够替代工程师进行工程设

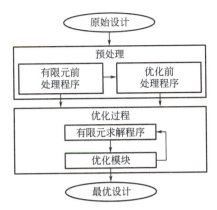

图 11.1 零部件形状优化设计过程

计的工具,但事实上不是这样,因为任何优化设计软件都不能确定应该优化哪些对象、哪些是设计变量、需要改变哪些量或参数,所以优化设计软件只是工程师进行设计的一种工具,其用途由其优化计算的能力决定。

优化设计软件通常要进行零部件几何外形的优化计算,一般要具有较强的数值计算能力。大多数零部件结构优化设计的问题都可以看作数学中的极值问题。求极值的有效方法很多,但方法太多也不好,因为对于一个特定的问题,其最佳解决方法只有一种。利用优化设计软件可以很好地解决这个问题,因为优化设计软件不仅可以帮助工程师选择解决问题的方法,而且能够帮助工程师找到最佳解决方法。

我们经常需要优化零部件或装配体在实际工作过程中承受的最大应力,所涉及的设计变量一般是零部件或装配体的几何尺寸,如指定零件的厚度。一般首先用 CAD 系统构建零部件或装配体的几何外形,如果设计结构正确,那么工程师会选择相应的有限元分析软件,对上述结构的机械性能进行数值模拟;然后根据计算结果(如最大应力的分布状况)判断设计是否有效。在设计过程中,工程师可能需要改变 CAD 或 FEA 模型的一些参数或特征属性,如零部件或装配体的几何尺寸、材料参数及约束和加载的状况。例如,在 CAD 系统中改变某个零件的厚度或增加一个孔,它的有限元分析模型会自动进行相应修改。在大多数情况下,工程师采用线性静力学的方法分析应力状况,这种方法的优点在于能用较少耗时、较多有限元分析单元得到需要的有限元分析结果。但该方法也存在缺点,例如估算处于运动状态的零部件或装配体的载荷大小或方向时,往往需要较丰富的专业知识,而这种方法无法满足要求。

2. 优化设计的数学理论方法

利用有限元分析方法进行优化设计的过程一般有以下三个步骤。

(1) 在 CAD 系统中构造出某零部件或装配体的几何模型。

(2) 建立相应的有限元分析模型。

(3) 对有限元分析的计算结果进行分析和判断。

一般人工优化设计过程除了涉及上述三个步骤,还要根据计算结果来判断优化设计的合理性;如果设计结果不合理,就要修改步骤(1)、步骤(2)。有限元分析的结果就是优化的结果,每个输入 CAD 或 FEA 模型的参数或特征属性都可以看作设计变量。优化设计算法对许多有限元分析都有指导作用,它的每种算法都会对不同的设计变量产生不同的数据组，所以 CAD 系统与 FEA 系统必须具有关联性，以将人工优化设计转化为优化设计算法。例如，开始对某零部件或装配体进行有限元分析时，工程师一般要对其有限元分析模型上的某平面施加约束，并假定这个平面就是零部件或装配体 CAD 模型上的平面。优化这个平面的结构时，其 FEA 模型上的平面会随 CAD 模型的变化而变化，这样上述约束就会施加在改变后的平面模型上。为了实现 CAD 模型与 FEA 模型的自动数据交换，CAD 系统与 FEA 系统必须具备这种关联性。在定义了机械系统的优化设计问题之后，就可以通过数学描述来解决这类问题。

大多数优化设计都要解决以下三个基本问题。

（1）目标函数的最小值（或最大值）。例如，设计汽车仪表板时，需要其在某指定区域上受到的应力最小。

（2）影响目标函数值的设计变量组。例如，设计汽车仪表板时，需要确定仪表板几何外形和材料的变量。

（3）约束条件。约束条件使优化设计中的变量只能在某范围内取值。例如，设计汽车仪表板时，需要限制其质量。

实际上，建立一个无约束条件的优化问题也是非常可能的。也许有人会认为几乎所有问题都应具有一定的约束条件。例如，汽车仪表板的厚度不能为负值，但实际上无须对一些设计变量施加约束条件，也可以获得与基本常识相符的结果，如上述汽车仪表板厚度为正值的问题。

3. 优化设计的优点和缺点

许多应用软件都以解除或减少人的重复性工作为目的。基于计算机的优化设计技术属于一种最新的应用设计技术，其目的就是增大计算机的计算量，以减少人的工作时间。实际上，进行优化设计计算时，使用计算机需要的计算量甚至比人工设计方法。因为优化设计技术采用了严谨的数学计算方法，所以它的设计效率要比人工设计效率高。当然，基于计算机的优化设计技术取代不了人的思维，因为人的思维有时可以大大缩短设计过程。基于计算机的优化设计方法与人工方法相比，其明显优点是如果优化设计软件使用正确，它就能够考虑所有设计方案。也就是说，它会考虑各种可行的设计参数，因此利用优化设计软件进行计算的结果应是最精确的。

第 12 课　工程机械开发中的动态仿真技术应用

所有工程师可能都了解虚拟样机技术的应用目的。现在出台的一些严格的法律制度（如关于废气排放和噪声）和较高的用户指标（如有关设备性能和操作方面）均需要更先进、更复杂的技术来完成产品设计。如果采用传统设计方法，则需要更多的研制费用和较长的研制周期；但竞争激烈的市场要求降低产品的研制费用并缩短产品的研制周期。

汽车行业已经广泛采用虚拟样机技术，目的是解决行业中出现的上述矛盾。人们已经能够对整车的各子系统进行仿真计算；但现在需要研究如何对车辆的整体性能进行数值计算，这是虚拟样机技术的发展趋势。沃尔沃 L220E 型装载机的多系统模型如图 12.1 所示。对整车进行数值计算的目的是评估车辆的控制性能、舒适程度和稳定性，以及测试车辆的碰撞性能。

图12.1 沃尔沃L220E型装载机的多系统模型

在汽车工业的设计领域中,越野行业相对滞后的原因主要体现在以下几个方面。

(1) 行业的规模较小。

(2) 购买最新CAE软件和培训上的费用太高。

(3) 制造的产品在外形和所使用的子系统上各不相同(这可能是最重要的因素)。

最近公布了一些对整车进行仿真计算的应用实例,涉及仿真技术、车辆子系统、舒适度和稳定性。本课讨论的是整车的动态仿真问题,目的是分析、优化车辆的整体工作性能。虽然研究对象是液压驱动式挖掘机,但得到的结果适用于其他越野机械设备。

1. 设计过程与可视化

在越野机械设计领域,人们研究的课题通常是如何变革现在使用的产品研制方法和设计过程,主要研究的问题是如何利用动态仿真技术改变现在使用的设计理念。因此,本课重点讨论如何利用动态仿真技术分析并优化越野机械设备的整体工作性能,这在前面已经提到过。

变革现在设计产品的总体目标如下。

(1) 较高的工作性能、生产效率和可操作性。

(2) 较好的越野性。

(3) 较短的研制周期。

(4) 较低的研制费用。

在越野汽车设计的早期阶段,人们已经有一些有价值的设计经验,即在越野汽车的反复改进设计阶段,尤其在其概念设计阶段,汽车速度设计非常重要。沃尔沃公司使用较先进的设计计算程序替代以前使用的程序,使得设计工作可以由专门的仿真计算程序完成。该程序是在多体动力学和现代数据库的基础上研制的。现在人们已经证明:对用户来说,这种新研制的仿真计算程序的适应性、精度和效率都比较高;但对从事预测性研究的工程师来说,其优越性并不明显,因为这种程序利用的是多体动力学仿真技术,而不是数学程序,所以每次运行的时间都比以往的程序慢2s。使用以往的程序优化越野汽车的受力状况非常耗时,使用新研制的仿真计算程序时,如果事先了解车辆的受力情况,并在设计时加以考虑,人们就完全可以研制出一个结构更为紧凑的程序来解决上述问题。虽然该程序的计算精度低,但速度比以前提高了。引进该仿真计算程序的主要目的是让从事预测性研究的工程师放弃使用计算时间短但精度不高的程序。

在课题进行过程中还发现了其他类似问题：目前进行初始静载荷计算程序速度较高，所得到精度也较高（图12.2）；但缺点是在研究整车的动态特性时，首先要进行概念样机测试，然后对真正的样机进行试验，最后改进设计。

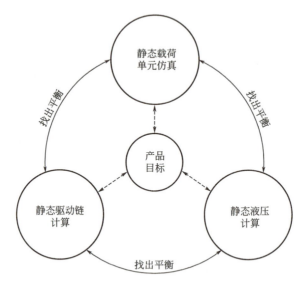

图12.2　初始静态载荷计算程序示意

与实际设计过程一样，在利用动力学仿真程序进行越野车辆设计时，首先将产品的设计目标函数值输入初始静载荷计算的程序，如果对计算结果不满意就修改目标值；然后进行所谓的"动态增加，静态计算"。只有计算误差在允许的范围内，程序才会对目标值进行迭代循环计算。由于进行迭代的目标值不可能包括设计对象的所有条件（包括动态特性），因此需要在一定范围内设置循环计算中的误差检验准则。只有当计算误差过大时才使用新的静态计算循环程序（需要重新输入产品目标值）。如果计算结果仍不合理，那么应修改目标值。由于整车的仿真计算需要较长时间，因此快速检验初始循环计算结果非常必要。通常，检验过程在循环计算的第二步完成，但具体方法仍需要改进。比如，在挖掘机性能的循环计算中存在临界阶段，使装满货物的机械设备（该设备带有装满货物的大桶，如挖掘机）改变运送货物的方向，此时车辆的各子系统存在相互协调的关系。

（1）为了挖掘机转向，操作人员往往要降低发动机转速（否则齿轮换挡速度较高可能会使传动联合器提前失效、损坏）。发动机转速降低了，其转动所需转矩也会相应变小，但由于涡轮增压器具有转动惯性，因此发动机的起动变得非常迟缓。

（2）当换向齿轮由反转变为正转时，由于此时挖掘机仍在后退，因此发动机的涡轮旋转方向发生突变，涡轮齿轮的轮齿间发生相对滑动，造成发动机需要的转矩急剧增大。

（3）有些操作人员在挖掘机转向时不停止装载货物，这样就需要更多的油液提供动力。通常发动机需要的油液量和液压泵的排液量、发动机机轴的转速成一定比例关系。假定液压泵与发动机机轴直接相连，当发动机转速降低、需要的油液量增加时，液压泵的排液量也接近最大值。发动机液压泵运转需要的转矩与该泵的排液量、液压成一定比例关系，因为挖掘机装满货物（包括装载货物的大小）时，其液压泵的液压升高，排液量趋于最大值,所以需要的转矩也接近最大值。

当热效率降低的发动机上的驱动力和油泵液压力突然增大时,发动机是否能安全运行,取决于上述三种因素存在的时间长短,这可以对挖掘机进行仔细的动力学计算,得出满意的结果(对实际制造的功能性样机进行试验也可以得出满意的结果)。然而,最近似、耗时最少的计算方法如下:首先确定齿轮换向时的发动机转速,然后计算出油泵轮和转矩转换器涡轮之间相对滑移的最大量,最后可借助转矩转换器说明书上的参数求出发动机正常运转需要的最大转矩值。现在考虑另一种最不利的情形,假定液压力和油泵的排液量均达到最大值,已知发动机转速,可计算出发动机正常运转所需的转矩。如果这两次计算的转矩的和大于该速度发动机稳定运转的转矩,那么设计的系统方案在实际运行过程中可能会出现动力学平衡的问题;如果小于发动机稳定运转的转矩,则方案能按要求运行。为了研究上述两种情况的关系,有必要考虑发动机涡轮增压器的转动惯量和烟雾限制器等因素。由于正在加速的发动机很难获得低速稳定运转的转矩,因此只依据发动机静态转矩曲线得到的结论是错误的,此时发动机运转状态不安全。

现在我们需要研究发动机动态转矩。该图是在发动机加速度较低且静态转矩和动态转矩差别不太大时测量出来的(图12.3中所示的虚线)。

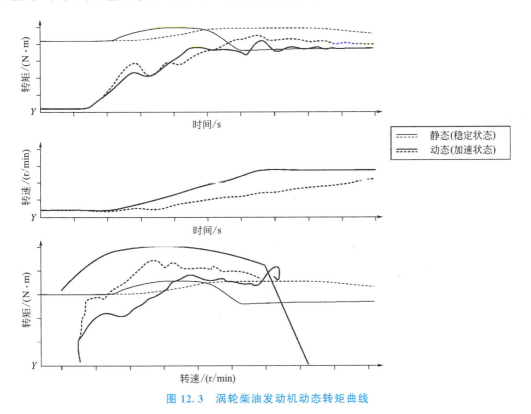

图 12.3　涡轮柴油发动机动态转矩曲线

假定车辆倒车时,发动机空转,此时发动机的加速度可以从其最大转矩时的速度和空转时的速度差中求出。利用带有电磁装置的试验台对发动机加速状况进行测试,测出的转矩就是制动需要的转矩。

在图 12.3 中,虚线代表的是发动机加速较低时的测量结果。从理论上讲,这种情况下的转矩应更大一些，原因如下：

（1）发动机旋转部件在加速时需要的转矩较小。

（2）发动机用于增压的时间较长。正如图12.3（a）描述的那样，发动机加速度增幅不大时，其转矩的增幅也不大。理论已经证明，发动机转矩的增大幅度与其速度密切相关，可以从发动机加速度增幅不大时的静态转矩仿真曲线、动态转矩仿真曲线的差别中看出来。如果发动机的加速度增幅较小，其动态转矩曲线（粗实线）会向静态转矩曲线（细实线）的方向延伸。

为了尽快找出有动力学问题的设计方案，需要研究更多检验方法，同时改进目前使用的一些方法和规律。

现在仔细研究这些经过改进的设计方法，即使这些方法在前文被称为"一种在原有基础上进行发展的设计方法，不是一种创新的设计方法"，也可以清楚地看出该设计方法与以往的方法相比变化很多。通过对该方法的介绍，我们明白建造功能性样机的目的是验证已经经过动力学仿真计算的方案是否可行；而进行动力学计算是为了确定某设计方案是否可行，是否需要重新设计。我们应该摒弃在实际应用中不可行的设计方案，只有经过实验检验的设计方案才可以进一步用于建造功能性样机，所以没有人会事先知道哪种设计方案是可行的。为了避免上述情况，应树立对产品的设计信心，处理好这两种方法的关系。

2. 总结和展望

Volvo挖掘机制造公司和林雪平大学（Linköping University）合作研究的科研项目的作用是通过对整个挖掘机进行仿真设计来分析和优化整机性能。在项目的研究过程中，企业的目的是在最短时间内利用最低成本研制出高性能、高效率和高操作性的产品。目前已经研究出一些经过改进的设计方法，同时制订了未来的研究计划。未来的研究重点将集中在确定整机操作性能，解决其实际使用过程中出现的动力学仿真问题。

从长远角度看，仍需研究更为复杂的设计方法和整机操作性能优化，当然也包含一些控制方法的问题。

第13课　机械加工中的热管技术

机械加工过程中产生的大部分能量在切削区都转化为热，造成刀具和工件的温度升高。过高的温度一方面会缩短刀具的使用寿命，另一方面会使工件发生热变形且加工的尺寸精度降低。此外，过高的温度会引起刀具尖端的边角变形。

在钻削过程中，钻头的温度特别高，因为吸收大量切削热的钻屑聚集在钻头与工件接触的狭小空间内。与其他切削过程相比，钻屑与钻头的接触时间较长。因而在相同的加工条件下，钻头的温度要高于其他机械加工刀具的温度。为了降低钻头的温度，通常采用向切削区浇注切削液的方法。目前使用的切削液有三种：添加硫黄、氯和磷的切削油，乳化液，以及人工合成的无机且具有降温作用的切削液。但在机械加工中使用的切削液会对环境和人类产生一定的毒害作用。例如，将切削液排放到土壤和水里，其中起润滑作用的化学物质会对周围环境产生毒害作用；操作人员经常接触到以液体或雾滴形式存在的切削液，切削液中的化学物质对他们也有极大的毒害作用。为了消除或减少切削液对环境和人类的毒害作用，研究人员提出了用低温液态氮和二氧化碳代替传统切削液的方法。尽管这种方法在延长刀具使用寿命和减少切削液用量等方面有一定的应用前景，但现在仍有许多技术问题没有解决，其中包括安全问题。近年来，机械加工行业出现了一个很有前景的发展趋势，即减少切削液的用量，因为相关研究表明，切削液是造成工业污染的主要原因之一。为了实现干式切削加工，人们希望采用一种有别于切削液的冷却方法来降低切削温度。

已经证实，热管是一种能够代替切削液实现高效传热的技术，它能够以干式或环保的方式将钻削过程中钻头上的切削热带走。热管的主要部件包括密封腔室（管壁和端盖）、安装在腔室内的毛绒芯和少量处于平衡状态的工作介质。图13.1所示为热管结构及带有热管结构的麻花钻。通常，热管分为三个区域：蒸发段、绝热段和冷凝段。从蒸发段输入的热量使热管内的工作介质汽化。蒸汽在汽化产生的蒸汽压作用下，经过绝热段扩散到达冷凝段并液化，释放热量。液化的工作介质在毛绒芯结构内的毛细压力作用下，流回到冷凝段。只要能保持足够的毛细压力，整个过程就会一直持续下去。

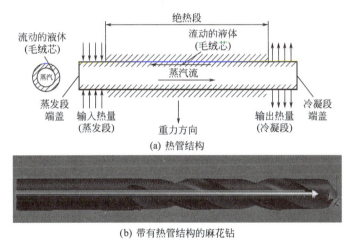

图 13.1 热管结构及带有热管结构的麻花钻

从传热学角度，研究人员已经做了大量研究工作以证明在钻削过程中采用热管技术是可行且有效的。主要发现如下。

（1）数值计算和初步实验结果均表明，在钻头内部使用热管能有效降低钻头的切削温度。麻花钻的切削温度如图13.2所示。

（2）麻花钻内的热管离钻头的尖部越近，其传热效率越高。虽然麻花钻上的热应力不随热管的长度增长而变大，但实际上加工条件限制了麻花钻内的热管长度。热管长度 l 与麻花钻长度 L 比值的影响如图13.3所示。

（3）热管直径对麻花钻钻头热应力的影响大于对温度的影响。虽然热管直径对钻头最高钻削温度的影响不大，但它对其上温度场分布状况的影响较大。热管直径 d 与麻花钻直径 D 比值的影响如图13.4所示。

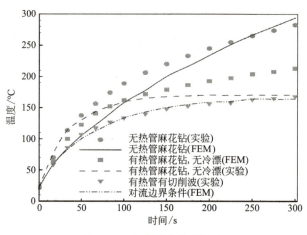

图 13.2 麻花钻的切削温度

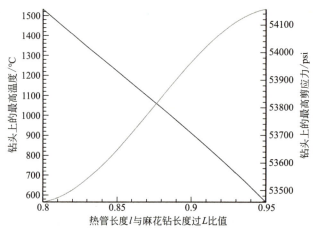

图 13.3 热管长度 l 与麻花钻长度 L 比值的影响

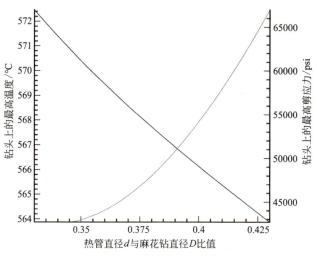

图 13.4 热管直径 d 与麻花钻直径 D 比值的影响

第 14 课　材料成形导论

1. 材料成形工艺体系

材料成形是将固体材料工件的原有形状变成另一种形状而不改变工件材料的质量和化学成分的一系列制造工艺方法的总称。

材料成形通常与制造工艺分类中的变形通用。机械制造工艺可以分为以下六大主要类别。

第Ⅰ类——最初的成形：将材料由熔融态、气态或未定型的固态颗粒制成一定的形状，即将材料的粒子牢固地聚合起来。

第Ⅱ类——变形：将固体材料由一种给定的形状转变为另一种形状而不改变原来的成分和质量，即保持聚合状态。

第Ⅲ类——分离：切削加工或材料的去除，即分离。

第Ⅳ类——连接：将单个工件连接到一起形成大的零部件、工件等，即使分散的工件连接起来，增大工件间的聚合力。

第Ⅴ类——涂覆：在工件表面涂覆薄层，如电镀、刷油漆、喷塑等，即使材料的基体与涂覆层产生聚合力。

第Ⅵ类——材料属性的变化：有目的地改变材料性能，以在加工处理过程中获得某些方面的最佳性能。这些方法包括改变微粒的取向，以及通过扩散产生或消除这些微粒，即重排、增加或减少微粒。

在机械制造工艺特别是在第Ⅰ～Ⅳ类工艺中，我们总是要面对在兼顾保证满足公差、表面构造及材料性能的前提下，如何最经济地制造一个特定的科技产品的问题。

2. 冷成形及热成形

对材料施加大于屈服强度的应力，可以使其产生形变硬化，或进行冷却加工的同时产生变形，完成这一变化需要采用许多制造工艺技术（如金属丝拉拔等）。利用冷加工及热加工的几种材料成形工艺方法如图 14.1 所示。

许多冷加工工艺方法都可用于材料成形并同时使金属强化。例如，轧制［图 14.1(a)］用于生产金属板材及薄板等。锻压［图 14.1(b)］是将工件材料置于模具型腔中，以获得形状相对复杂的零部件，如汽车的曲轴、连杆等。拉拔［图 14.1(c)］是将一根金属棒料经过拉拔模拉拔，使其变成线材或纤维。挤压［图 14.1(d)］时，材料被推入模具中挤压以获得横截面形状均匀的产品（如棒类、管类产品及门窗用铝合金镶边等）。拉深可用于制造铝合金饮料罐等。冲压［图 14.1(e)］、弯曲及其他工艺方法都可用于材料的成形。冷加工是一种在金属成形的同时又使金属强化的好方法。用冷加工成形可以获得尺寸公差小且表面质量好的零部件。值得一提的是，许多工艺方法（如轧制等）可以进行冷、热加工成形。

采用热加工方法也可使金属产生变形以获得所需形状。热变形被定义为金属材料在再结晶温度以上的塑性变形。热变形加工适用于较大零件的成形，因为金属在高温下的屈服强度低而塑性很好。另外，具有密排六方晶体结构的金属（如镁等）在高温下的滑移系增

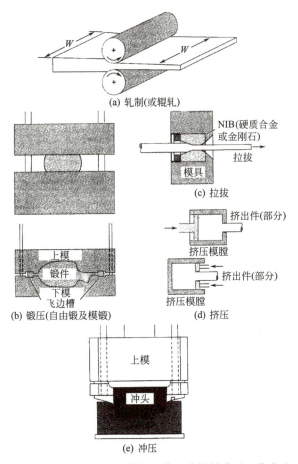

图 14.1 利用冷加工及热加工的几种材料成形工艺方法

加。高塑性使热成形允许的变形量比冷成形大。例如，一块厚板可以经过一系列连续热加工而变为薄片（板）。

热加工的优点是能消除材料中的缺陷：可以消除原材料中的一些缺陷或使其不良影响最小；气孔能被压合或熔合，材料中的成分偏析减少；通过再结晶可以很好地细化及控制金属的显微组织。因此，经过热加工的金属的机械性能及物理性能都得到明显提高。

3. 塑性成形

塑性成形是金属成形工艺方法数学处理的基础。材料科学及冶金学可以解释金属固体塑性状态的本质及影响参数，如成形速度、前期工艺历史及温度等。但早期的塑性成形理论主要用于计算应力、力及变形量。

塑性成形理论是基于宏观上观察到的现象建立起来的，换句话说，其以材料在变形过程中（如拉伸及挤压试验）所能观察到的或测试出的性能为基础，从而引出下面关于塑性的简单定义。

塑性是材料在外力（当应力状态达到该材料的临界值，即屈服强度或初始流变应力时）作用下永久改变形状的能力。正如我们在拉伸试验中看到的那样，当应力低于屈服强度时，去除载荷后变形将自行回复。此时，材料的变形为弹性行为。如果应力增大到高于屈服强度，则会产生永久变形，卸载应力后，工件形状与原来不同。此时，我们说材料发生了塑性变形或永久变形，或者最终变成了一定的形状，即材料成形了。具有弹、塑性行为的材料在产生永久变形后再次承载时，在达到流变应力前不会发生塑性变形。这种由前面的变形导致流变应力增大的现象称为加工硬化或形变硬化。

加工硬化可以通过回复和再结晶的动态软化过程消除。产生了加工硬化的显微结构由含有大量复杂的位错变形的晶粒组成，将这种金属加热时，热能会促使位错运动并形成多边形亚晶粒的边界。但是，位错密度不会变小。这种在较低温度下消除残余应力而没有改变位错密度的过程称为回复。

当把产生了加工硬化的金属材料加热到再结晶温度以上时，快速的回复将进一步消除残余应力并形成多边形的位错结构。新的细小的晶粒在多边形晶粒的晶界处形成核，同时大多数位错被消除。由于位错数量大大减少，因此再结晶金属的强度下降而塑性增大。这种通过对冷加工材料进行热处理以形成新晶粒的过程称为再结晶。再结晶时如果加热温度太高，则会引起晶粒变大。

4. 材料成形方法

材料成形方法可分为五种，主要是根据实际有效的应力的显著不同分类的。由于有不同的应力状态同时存在，或者在变形的过程中应力状态可能改变，因此根据工艺的种类将应力状态简单描述为某种形式是不可能的。因此，只能选择最主要的应力状态作为分类依据，从而将材料成形方法分类如下。

（1）压缩（压制）成形（在压应力作用下成形）：这种固态形变为塑性状态主要是由单向或多向压缩载荷的作用而获得的。

（2）拉压成形（在拉应力、压应力结合的载荷作用下成形）。

（3）拉伸成形（在拉应力作用下成形）。

（4）弯曲成形（借助弯曲应力的作用成形）。

（5）剪切成形（在剪应力作用下成形）。

第15课　材料成形工艺

本课通过一些实例简单介绍几种材料成形工艺，但不包括装配和连接工艺。

1. 锻造

锻造是对固态金属材料进行的最基本的机械塑性成形加工工艺方法，整个加工过程遵循质量守恒定律。从工艺学上来说，锻造可定义为在冲击力或压力的作用下，经过一定的塑性变形使金属通过成形、精锻及力学性能的提高而提高其使用性能的工艺方法。锻造有很多类型，图15.1(a) 所示为最普通的落锤锻造：金属加热到适当的温度后，将其放入型腔；上型腔与下型腔合拢，迫使金属充满型腔；多余材料挤入型腔周边的飞边槽内，由后续的切边精整工艺清除。提到锻造，通常是指热锻，因为冷锻有其专门的名称。锻造工艺中损失的材料一般很少。由于锻造的公差和表面粗糙度通常不能满足产品的设计需要，因此需要对锻造的零件进一步加工。锻造机械包括落锤、机械驱动或液压驱动的锻压机。这些机械设备一般只做简单的平动。

2. 轧制

轧制是对固态金属材料进行的最基本的机械塑性成形加工工艺方法，整个加工过程遵循质量守恒定律。轧制广泛应用在板材、薄板和结构桁条等制造中。图 15.1(b) 所示为板材轧制。铸造生产出的铁锭加热后，经过加工，其厚度变薄。但由于工件的宽度保持不变，因此其长度将随着厚度的变薄而变长。热轧的最终阶段是冷却，这样可以提高工件的表面质量、公差和强度。在轧制工艺中，根据需要通常将轧辊的外形设计成所期望的几何形状。

3. 粉末（压制）成形

粉末（压制）成形是对固态金属材料进行的最基本的机械塑性成形加工工艺方法，整个加工过程遵循质量守恒定律。这里仅介绍金属粉末（压制）成形工艺，一般成形砂、陶瓷材料的挤压等也属于此类材料成形工艺。金属粉末（压制）成形时，型腔内充满标称体积粉末，如图 15.1(c) 所示，图中施加了约 500 N/mm^2 的压力来压紧粉末。在成形过程中，粉末颗粒充满型腔并产生塑性变形。挤压后的密度通常是固态材料密度的 80%。塑性变形后，粉末颗粒"焊合"到一起，其强度足以满足一般的操作要求。挤压后，一般要对零件进行烧结热处理，该温度是材料熔化温度的 70%～80%。一定要控制好烧结时的空气，以防止出现氧化现象。根据材料和工艺参数，烧结过程的时间是 0.5～2h，烧结后的零件强度非常接近相应固体材料的强度。

闭合的型腔形状与最后得到的零件几何形状对应。粉末成形压机包括机械式压机和液压式压机两种，每分钟可生产 6～100 个零件。

4. 铸造

铸造是将液态材料充满型腔的最基本的机械塑性成形加工工艺方法，整个加工过程遵循质量守恒定律。铸造不但是最古老的加工方法之一，而且是人们最熟悉的材料成形工艺。材料被熔化并浇注到与铸件形状和尺寸相应的铸型型腔内 [图 15.1(d)]。液态材料充满型腔，随后冷却凝固，最终得到所需零件几何形状。

铸造工艺的流程包括制作适当的铸型、熔化材料、将材料充满或灌注进型腔内及冷却凝固。采用不同的铸型材料可以得到不同特性和不同尺寸精度的铸件。在铸造工艺中使用的设备有熔炉、铸型制作机械和铸造机械。

5. 车削

车削 [图 15.2(a)] 是对固态金属材料进行切除的最基本的机械加工工艺方法。车削工艺广为人知，使用最为普遍。在车削工艺中，人们使用切削刀具从工件上切除材料，形成切屑碎片，一般用于加工圆柱形工件。车削时，工件旋转，切削刀具纵向进给。与工件材料相比，切削刀具材料具有很高的硬度和耐磨性。车削时，可以根据需要采用不同类型的车床。车床通常由电动机驱动，通过不同的齿轮系向工件提供必要的转矩，使刀具完成进给运动。

利用相同的金属切削原理也可以得到其他完全不同的加工操作方法或加工工艺，钻削和铣削就是其中最常用的两种工艺方法，分别如图 15.2(b) 和图 15.2(c) 所示。由于使用了不同的加工刀具，因此其采用的操作方法完全不同。使用不同形状的刀具并使刀具相对于工件进行不同形式的运动可加工出不同形状的零件。

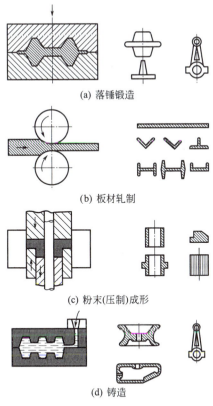

(a) 落锤锻造

(b) 板材轧制

(c) 粉末(压制)成形

(d) 铸造

图 15.1　材料成形工艺

6. 电火花加工

电火花加工（electrical discharge machining，EDM）是通过熔化和蒸发，对固态金属材料进行以热变化为主要方式的加工工艺方法，如图 15.2(d) 所示。电火花加工时，工件和工具（电极）之间产生了许多电火花，这些电火花对工件材料有侵蚀作用，从而达到去除材料的目的。这时工具上常常留有工件外形的反转形状。当工件和工具之间的电压差足够大时，其间的流体介质被击穿，在电压作用下，流体介质进入工具和工件之间的缝隙，形成了传导火花的通道，产生放电现象。流体介质一般是矿物油或煤油，它的作用是作为绝缘的流体和散热剂，保持均匀的电阻并带走从工件上蚀除的材料。电火花通常以每秒成千上万次的速度出现，一般积聚在工具和工件之间最小缝隙的点上，因而产生了大量热量，这些热量将工件的材料侵蚀并散入液体中。电火花加工的材料表面特点是其上有许多小蚀坑。

7. 电解加工

电解加工（electrochemical machining，ECM）是通过电解分解对固态金属材料进行以热变化为主要方式的加工工艺方法，如图 15.2(e) 所示。工件的电解是通过电路来实现的，工件为阳极，工具为阴极，工具上留有工件外形的反转形状。电解液通常使用水基盐（10% 的氯化钠和 30% 的硝酸钠），电压一般为 5～20V，可维持的电流密度为 0.5～2A/mm^2，移动速度为 0.5～6cm^3/(min·1000A)，但上述的这些参数通常需要根据具体的工件材料确定。

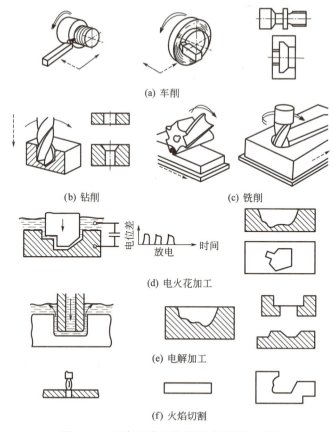

图 15.2　固态下的工件材料质量减少工艺

8. 火焰切割

火焰切割是通过燃烧对固态金属材料进行以化学变化为主要方式的加工工艺方法，如图 15.2(f) 所示。火焰切割前，一般需要将材料（含铁金属）加热到可以与氧气发生燃烧反应的温度。理论上，燃烧释放出的热量足够维持上述反应；但是由于部分热量散失到空气或传递到材料中，因此要持续地提供热量。人们目前使用喷枪来提供上述反应所需的热量，使用最广泛的是乙炔切削喷枪，它通过燃烧乙炔和氧气发生的化学反应产生热量。切削用的氧气通常由喷枪顶端中心的孔提供。

火焰切割工艺仅适用于易燃材料。对于其他材料，可采用以热变化为基础的切割工艺（如等离子切割等）。

9. 精密冲裁

精密冲裁（简称精冲）是一种用来生产冲裁件的工艺。冲裁件往往表面平整且有光洁的切边，其精度可以与精加工后的零件媲美。当需要加工大批零件，费用合理，省去刮削、修整等工序时，应采用这种快捷、简便的工艺方法。

有一种精冲工艺方法使用圆形刃口模具和小间隙进行冲裁。这种工艺适用于落料，但冲孔效果不太理想。在这种工艺中，模具的刃口半径需要根据加工材料的硬度、厚度及零件的具体形状选定。精冲的基本特点如下：在实际加工过程中，模具的刃口半径往往取最小值，从而得到质量较高的零部件；根据情况，最小半径一般为0.3～2mm。

设计精冲模时，冲头与下模之间的间隙总是比一般冲裁模具的间隙小。通常认为，模具的总间隙为0.01～0.03mm时，可以获得质量较好的精冲件。需要强调的是，这里的总间隙不是孔或者落料件的单边间隙。

第16课　模　具　导　论

模具是工业生产的基本技术装备。人们借助模具完成工业产品的成形，并且这些模具是为产品特别设计和制造的。因为模具的型腔形成了零件的形状，所以模具是零件生产过程的核心。模具类型很多，有铸模、锻造模、陶瓷模、压铸模、拉丝模、玻璃模、磁铁成形模、金属挤压成形模、塑料模、橡胶模、塑料挤压模、粉末冶金模、冲压模等。

下面介绍一些模具成形工艺及其模具。

1. 冲压成形

冲压成形是最不复杂的模具成形工艺，适用于大型零件或小批量生产。对于小批量的生产需求，制造冲压模具要比制造注塑模具经济得多。冲压模具常用作初始制模，所制作的试样用于组合件的试装配和组装，从而在制造用于大批量生产的注塑模具之前对冲压模具进行改进。

冲压成形适用于对公差要求不高的设计。冲压模具结构简单，可以由两块带有型腔的模板组成；也可以由一块带有型腔的模板，另一块不带有型腔的模板组成。此外，可以在上、下模板之间加一块板子，以便在成形部件内形成孔穴。图16.1所示为两模板单型腔冲压模具，该冲压模具无需加热元器件和温度控制器，它的成形温度完全由成形压力控制。

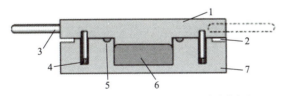

1—顶板；2—开式条形槽；3—手柄；4—定位销及衬套；5—溢料和裂缝修整浇口；6—零件型腔；7—底板。

图16.1　两模板单型腔冲压模具

由于冲压模具结构简单，因此它是市场上价格较低的模具，广泛应用在小批量的产品制造中。

2. 注塑成形

注塑成形是较复杂的成形工艺。由于注塑模具结构设计复杂，因此购买注塑模具的成本比铸造模具和冲压模具高。虽然注塑模具成本较高，但其循环周期比其他工艺短，零件的成本也比较低，尤其是在工艺过程实现自动化的情况下。注塑成形适用于外形精致的零件成形，它可以在材料上产生高达29000psi的压力，将材料填充到模具型腔的各个角落。

注塑成形中采用的模具由两部分组成：一部分是静止的，另一部分是可移动的。静止的部分直接固定在固定模板上，在操作时与注塑装置的喷嘴直接接触。可移动的部分固定在活动模板上，通常装有脱模装置。采用这种平衡流道系统的模具可以将塑料从浇口送到各型腔中。

注塑模具可以是由两块模板组成的结构简单的模具，这种模具有浇道系统，可以将橡胶复合物从合模线处注入各型腔；也可以是由许多块模板组成的结构复杂的模具，这种模具的模芯内有脱模销和附加的加热元件。

图 16.2 所示为三模板多型腔注塑模具和两模板多型腔注塑模具的基本结构。不要求这些模具有加热元件或温度控制器，其成形温度完全由所施加的注塑压力控制。

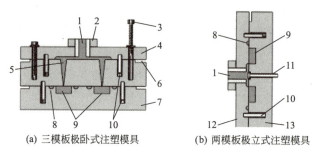

(a) 三模板极卧式注塑模具　　(b) 两模板极立式注塑模具

1—注塑浇道；2—喷嘴套筒；3—脱模螺栓；4—顶板；5—浇口；6—开式条形槽；7—底板；8—溢料和裂缝修整浇口；9—零件型腔；10—定位销及衬套；11—脱模销；12—固定板；13—移动板。

图 16.2　三模板多型腔注塑模具和两模板多型腔注塑模具的基本结构

3. 铸造成形

铸造成形有两种方法：开式铸造成形法和压力铸造成形法。采用开式铸造成形法时，先将液体混合物倒入敞开的模具型腔，再使其固化；采用压力铸造成形法时，先将液体混合物倒入敞开的模具型腔，再将盖子盖好，对型腔加压。压力铸造成形法常用于结构复杂的零件或泡沫材料的成形。

铸造成形的原理与注塑成形的原理相似，但使用的成形材料不同。事实上，铸造成形可产生与注塑成形具有相同几何形状的零件。在许多场合，由于零件的成本低，因此通常用注塑成形替代铸造成形。但对于结构较特殊的零件，尤其是壁厚较小的零件，铸造成形通常是更好的选择。

由于模具的材料通常是以低压液体形式流动的材料，因此其制造费用往往不是很高，适用于小批量生产和样模制作中，不适用于中等批量生产。用聚氨酯制作的液体印模具有极强的抗磨蚀、抗冲击和抗弯曲疲劳的能力，而且可用于制造较复杂的零件形状和较厚的零件横截面。但与其他成形工艺相比，这种工艺的循环周期和固化时间都比较长，并且材料一旦成形，就不能对其进行重新研磨和使用。

铸造成形模具的结构与注塑成形模具的结构几乎完全相同。它主要由脱模部分和顶盖部分组成，两者在合模线处会合。通常型腔和模芯均加工在上述两部分内的嵌件上。模具的顶盖部分固定在静止模板上，脱模部分固定在活动模板上，型腔和加工模芯的设计必须方便模具从铸件的位置脱离。当模具打开时，要求用脱模销将零件从模具中移出，同时为了防止零件和模具之间黏结，必须在型腔内喷射润滑剂。

模具通常由工具钢、模具钢或马氏体时效钢等材料制成。由于模具材料上没有天然的透气孔,并且熔化的金属注入时会快速流入模具内,因此在合模线处必须开设排气孔和通道,以便型腔内的气体排出。

4. 挤压成形

虽然挤压模具的结构相当简单,但是挤压成形工艺在制定和加工过程中要非常仔细,以保证产品的设计与制造一致。压力是通过有一定外形的模板施加的,但需要注意控制进给速率、温度及压力的变化幅度。

与冲压成形或注塑成形不同的地方是,橡胶从挤压模具中脱落时尚未固化。未加工的橡胶只有放置在圆盘或长盘(取决于其外形)上,并且装入高压釜后,在一定的热量和压力的作用下才能固化。

较长的橡胶可采用盐浴固化系统成形,而硅橡胶可采用连续加热装置成形。采用的固化工艺方法取决于要求挤压模具的数量和外形。

大多数挤压模具其实就是简单的圆柱形钢件,但其内部需要有起挤压作用的轮廓外形,同时必须留有一定的余量,以便成形复合物的收缩或膨胀。挤压模具是结构较简单的模具。

虽然这些模具建造起来相对简单,但是由于涉及的工艺不同,因此哪怕是最少的浇注过程,其外形都可能会有所变化。

第 17 课　模具设计与制造

目前 CAD 和 CAM 技术已经广泛应用在模具的设计和制造中。例如,首先利用 CAD 软件在计算机上构建出模具的模型,然后采用三维动画的方法从各角度观察模具的结构,最后将模具的各种参数(如压力、温度、冲力等)导入数字模拟模型并进行模拟试验与分析。CAM 能够控制模具的制造质量。采用上述计算机技术对模具进行设计和制造有很多优点:较短的设计时间(随计算机的运行速度变化)、较低的制造成本和较高的制造效率等。这种新的设计、制造方法适用于小批量的模具生产,可以在最后时刻对某个特定的模具零件进行修改。此外,这些新的工艺过程可以用于制造复杂的模具零件。

1. 模具的计算机辅助设计

通常手工绘制模具结构图是一项费时的任务,虽然它不属于创造性工艺过程,但对工艺过程来说必不可少。

CAD 采用计算机及其辅助装置简化设计任务和提高设计质量。CAD 系统提供了一套高效的设计方法,它可以和坐标测量等其他检验设备一起使用,组成模具设计的检验程序。选择工艺顺序时,CAD 系统通常能够发挥十分重要的作用。

CAD 系统由三个基本部分组成:硬件、软件、用户。CAD 系统的硬件部分包括处理器、系统显示器、键盘、数字转换器和绘图仪;软件部分由能够完成设计和画图功能的程序组成;用户是模具的设计者,采用 CAD 系统的硬件和软件来完成模具的设计过程。

根据产品的三维数据,首先需设计模芯和型腔。通常设计人员可以对模芯和型腔的相关零件进行预设计,这说明现在设计模芯和型腔的方法改变了。目前 CAD 系统可以进行如下设计过程:首先计算分模线,将零件分成模芯和型腔,生成流表面和截流表面,根据计算出的最佳零件草图确定出型腔、滑道及嵌件的位置和方向。然后在概念设计阶段，粗略确定模具部件的位置和几何形状，如滑动装置、喷出系统等。此时可根据上述设计数据确定模芯和型腔板的尺寸，并从产品标准目录中选取标准模具。如果没有一个标准模具与上述设计要求符合，则可以选择离要求最接近的标准模具，并作相应修改，如通过调整约束条件和结构参数的方法使某尺寸的板子可以在设计中应用。最后，对功能部件进行细化设计，并添加一些标准部件，完成整个模具的结构设计，整个过程是在三维空间中完成的（图17.1）。此外，模具系统提供了检查、修改和细化零件的功能，因此，在模具设计的早期阶段就可以自动生成图纸和材料清单。

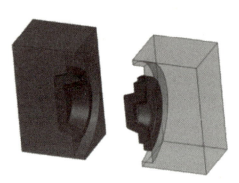

图 17.1　模具的三维立体模型

运用模具设计系统的三维结构设计方法和智能化功能，可以在模具开始设计阶段避免二维设计中出现的典型错误，如冷却系统和部件/型腔结构之间的干涉及孔的定位错误等；也可以在设计的任何阶段生成零件图样和材料清单，从而及时订购模具材料，并维护最新文档以便与客户进行讨论或为模具制造企业提供投标参考。

使用专门的三维模具设计系统能够缩短模具的研发周期、提高模具的设计质量、增进团队合作精神、使设计人员从单调乏味的日常工作中解脱出来。然而经济上的成功主要取决于合理规划的工作流程，只有采取了适当的组织方法和人员考核制度，才能缩短产品的研发周期。因此，零件设计、模具设计、电气设计及模具制造等部门必须紧密合作、协同工作。

2. 模具的计算机辅助制造

一种能有效减少制造费用、缩短研发周期的方法是建立能够充分发挥设备和人员潜能的制造系统。这类制造系统的基础是 CAD 系统设计模具的数据。这些数据可以帮助我们对主要工艺过程作出决策，提高模具制造的精度，减少非生产时间，这个过程称为 CAM。CAM 的目的是在可能的条件下，利用计算工作站的计算机直接生产出模具部件而不需要任何中间步骤。

CAM 系统的自动化程度不仅体现在零件某个特征的制造上，而且体现在该零件所有细节特征的制造上，从而最终实现制造路径的最优化。当需要制造多种特征时，CAM 系统会为制造者制订一个工艺规划，相关操作都要依据这个规划，目的是减少加工过程中变换刀具的次数和使用的刀具。

目前 CAM 的发展趋势是研制一些 CAM 方面的新技术和新工艺，如微研磨技术。该技术采用复杂的三维结构设计方法和表面制造精度高的技术制造出精度较高的注塑模具。CAM 系统将继续在智能化加工的深度和广度上发展，直至 CNC 编程工艺过程实现完全自动化。这种需求对于能够实现柔性化、组合机械加工的先进多功能 CNC 机床来说更是如此。CAM 系统将在机械工的操作下，使冗余的制造过程逐渐实现自动化，使其在计算机的帮助下更快、更精确地制造出产品。

在保证模具生产质量且强调生产高效的今天，模具制造商需要利用最新的软件制造技术快速地制订出工艺规划，制造出复杂的模具产品，并减少相应的生产时间。总之，模具制造业正朝着提高 CAD、CAM、CNC 数据交换质量的方向发展，同时 CAM 系统在加工过程方面变得越来越智能化，减少了模具产品的生产周期和总的加工时间。此外，五轴联动加工模式也可用在模具制造方面，尤其在进行深型腔加工的场合。随着电子数据处理（electronic data processing，EDP）技术在模具制造业中的应用，模具制造技术已经出现了新的发展时机，这将极大地缩短模具的生产周期、降低模具的生产成本，提高模具的生产质量。

第 18 课　金属热处理

普遍认同的对金属及合金热处理的定义如下：以一定的方式加热或冷却固态金属或合金，以达到一定的条件并获得某些性能。以热加工（如锻造）为目的的加热不在此定义之内。同样，有时用于生产诸如玻璃或塑料制品的热处理也不属于该定义的范畴。

1. 时间-温度-转变图

热处理的基础是时间-温度-转变图，即 TTT 图，三个参数都绘制在一个图中。

为了绘制 TTT 图，将特定的钢置于给定温度下，以预先确定的时间间隔检查其结构，记录发生转变的量。已知共析钢（T8）在平衡条件下，在 727℃ 以上时全部为奥氏体，低于此温度时为珠光体。为了形成珠光体，碳原子将扩散形成渗碳体。扩散是一种渐进过程，需要足够的时间完成奥氏体向珠光体的转变。对于不同的样品，可以记录下在任一温度下发生转变的量，然后把这些点绘制在一条以时间和温度为坐标轴的曲线上。通过这些点可以得到图 18.1 所示的共析钢（T8）的 TTT 图。左边的曲线表示在任一给定温度下

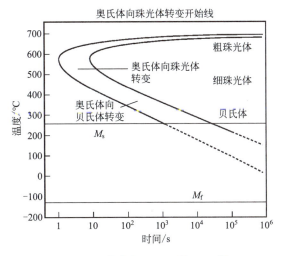

图 18.1　共析钢（T8）的 TTT 图

奥氏体开始转变为珠光体所需的时间。类似地，右边的曲线表示转变完成所需的时间。两条曲线之间是表示部分转变的点。两条水平线 M_s 和 M_f 分别表示马氏体转变的开始和结束。

2. 热处理工艺的分类

在一些场合，热处理过程可以根据工艺和应用明确地区分开来。而在另一些场合，因为相同的工艺常常可以用来达到不同的目的，所以对该术语的说明和简单的解释是不够的。例如，消除应力的热处理和回火处理常常使用相同的设备、相同的时间和温度循环实现，但这两种工艺的目的是不同的。

下面讲解热处理工艺及其相互关系。

（1）正火。正火是把铁类合金加热到指定的相变温度上限以上的合适温度，接下来在不流动的空气中冷却到至少比相变温度低得多的某温度的热处理工艺。对于低碳钢，得到的结构和性能与完全退火相同。而对于大多数铁类合金，正火和退火意义不同。

正火一般作为起调节作用的热处理工艺，特别是用于细化在锻造或其他热加工工艺中经受了高温的钢的晶粒。正火处理之后通常还要接着进行下一个热处理操作，如为了后续淬火、退火、回火而进行的奥氏体化。

（2）退火。退火是把合金加热到一定温度并保温，然后以合适的速度冷却的热处理工艺。退火主要用于降低金属材料的硬度，同时在其他性能或显微组织方面产生一些理想的变化。这种变化的目的是提高机械加工性能或电性能、便于冷加工（称为中间退火）、增强工件的尺寸稳定性，但也不仅限于此。如果仅用于消除应力，就称为去应力退火，与应力消除热处理是同义词。

当"退火"这个术语用于铁类合金且无其他限制条件时，指的是完全退火。这种处理的原理是把合金加热到其相变温度以上，再进行循环冷却，以使硬度下降最多。冷却过程的变化范围较大，要看具体合金的成分和特性。

（3）淬火。淬火是把钢或合金浸没在液体或气体介质中，从奥氏体化温度下快速冷却的热处理工艺。常用的淬火介质有水、5％的氯化钠水溶液、5％的碱性水溶液、油、聚合物溶液或气体（常为空气或氮气）。对淬火介质的选择主要取决于材料的淬透性及被处理材料的质量（主要是淬火部分的厚度）。

上面列出的各种淬火介质的冷却能力有很大的区别。选择淬火介质时，最好避免选择比达到预期效果所需的冷却能力强的溶液，以尽量降低处理零件开裂或变形的可能性。

淬火包括直接淬火、喷雾淬火、高温淬火、分级淬火、局部淬火、喷射淬火和等温淬火。

（4）回火。在铁类合金的热处理中，回火指的是把奥氏体化和淬硬的钢或铁再加热到预先设定的相变温度下限以下的某温度（通常低于704℃）的热处理工艺。回火提供了获得各种综合机械性能的方法。用于淬硬钢的回火温度通常为150~200℃。不要把回火与中间退火或去应力退火混淆，即使三种处理方法的时间和温度循环可能相同，被处理的材料的状态和目的也可能不同。

（5）去应力退火。与回火相似，去应力退火是把钢和铁加热到相变温度下限以下的某温度的热处理工艺。对于有色金属，去应力退火温度可以稍高于室温，也可以达到几百摄氏度，取决于合金的种类和期望去除的应力。

去应力退火的主要目的是消除在模铸、轧制、机加工、焊接等过程中在工件中产生的应力。通常这一过程是把工件加热到预先设定的温度并保持足够长的时间来降低残留的应力（这是一个与时间和温度有关的工序），接下来以较低的速度冷却，以免产生新的应力。

第 19 课　数 控 系 统

数控系统是一种利用程序实现自动控制的系统，加工制造设备采用数控技术后可由数字、字符和符号等控制。这些数字、字符和符号等被编码成按一定格式定义的指令程序，用于特定的加工或工作。这些指令可以采用两种二进制编码的数字系统定义，即电子工业协会（EIA）代码和美国信息交换标准（ASCII）代码。一般来说，采用 ASCII 代码编码的机床控制系统不能接收采用 EIA 代码编码的指令，反之亦然。当然，这样的问题已经逐渐得到解决。数控加工制造已经广泛应用于几乎所有的金属加工机床，如车床、铣床、钻床、镗床、磨床、回转冲床、电火花线切割机及焊接机床，甚至弯管机也可采用数控技术。

1. 数控系统的组成

数控系统主要由以下三部分组成。

（1）程序指令。

程序指令由一条条详细指令组成，制造装备按要求执行这些指令。最常用的指令形式可以按要求使机床刀具主轴位于工作台某具体位置。工作台是用于固定加工零件的制造装备。许多更高级的指令还包括主轴速度选择、刀具选择及其他功能。

（2）加工控制单元。

加工控制单元包括用于阅读和解释程序指令并将其转换为机床刀具或其他制造装备的机械动作的电子和控制硬件。

（3）制造装备。

制造装备是一种进行金属加工的数控装备。在常用的数控领域中，制造装备用于进行机械制造。制造装备包括工作台、主轴、电动机及控制驱动单元。

2. 数控系统的类型

数控系统主要有两种类型：点对点数控系统和轮廓数控系统。

点对点数控系统也称位置数控系统，比轮廓数控系统简单，其主要原理是移动刀具或工件从一个程序控制点到另一个程序控制点；通常，像钻孔加工，每个点都可以通过数控程序中的指令进行控制。点对点数控系统适用于钻孔、沉孔加工、沉孔镗孔、铰孔和攻螺纹等。冲孔机床、点焊机和装配机床等也都采用点对点数控系统。

轮廓数控系统也称轮廓路径数控系统，定位和切割操作都是以不同的速度沿控制的路径进行的。由于刀具沿路径进行切削，因此刀具运动和速度的精确控制及同步性能是非常重要的。轮廓数控系统常用于车床、铣床、磨床、焊接机床和加工中心中。刀具沿路径的运动称为插补。许多种插补方法都可用于处理在轮廓数控系统中生成光滑的轮廓线时遇到的问题，较常用的方法有线性插补、圆形插补、螺旋形插补、抛物线插补和立方插补等。在所有插补方法中，路径控制以刀具的旋转中心为标准，在数控程序中对不同类型、不同直径的刀具，以及在加工过程中的不同刀具磨损量给予不同的补偿。

3. 数控系统的编程

数控系统的程序包括使数控机床进行操作和加工的一系列指令。数控程序可以由数控机床内部的程序库开发生成，也可以由外部采购获得。另外，数控程序可以通过手工编写，也可以进行计算机辅助编写。

数控程序包括一系列指令系统和命令系统。几何类指令用于定义刀具和工件之间的相对位置和运动；加工类指令用于定义主轴转速、进给速度、刀具转速等；传送类指令用于定义刀具或工作台的运动速度和插补类型等；开关类指令用于冷却液供给、主轴旋转、主轴旋转方向选择、换刀、工件进给、夹具开关等。第一个用于数控编程的数控编程语言是20世纪50年代由麻省理工学院开发的，被命名为自动编程工具（automatically programmed tools，APT）。

4. 分布式数字控制和计算机数字控制

数控技术在批量生产和小批加工中，无论是在技术上还是在商业上都取得了显著成功。目前，已经有两种数控系统得到了发展，介绍如下。

（1）分布式数字控制。

分布式数字控制（distribute numerical control，DNC）可以被定义为有许多台加工机床的制造系统，这些加工机床相互之间由一台计算机直接连接并进行实时控制。这样，在传统的数字控制技术中采用的磁带阅读器在直接数字控制中被取消，从而保证了制造系统的可靠性。不使用磁带阅读器，被加工的零件程序可从计算机的存储器中直接传送到进行加工的刀具上。从原理上讲，一台计算机可以控制100台加工机床甚至更多。DNC的计算机按要求为每个进行加工的刀具提供加工指令，当机床需要控制指令时，计算机可以立即将指令传送到机床上。

随着DNC技术和计算机技术的飞速发展，数字计算机尺寸和价格的大幅度降低，数字计算机的计算能力大大提高，大量传统的以硬件线路为基础的加工控制单元被以数字计算机为基础的数字控制单元取代。20世纪70年代使用小型计算机，随着计算机的进一步小型化，早期的小型计算机逐渐被现在的微型计算机取代。

（2）计算机数字控制。

计算机数字控制（CNC）使用专用的微型计算机作为加工控制单元。数字计算机都用于CNC和DNC，应注意二者的区别，可以从以下三个方面区分。

（1）DNC计算机是将指令数据发送到许多机床中或从许多机床中搜集数据，而CNC计算机每次只控制一台或多台机床。

（2）DNC计算机一般位于与机床有一定距离的位置，而CNC计算机一般位于与机床较近的位置。

（3）DNC计算机开发的软件不仅用于控制产品的单件生产，而且用于企业制造部门的信息管理系统；而CNC计算机开发的软件一般只用于某特殊加工的工具。

第 20 课　虚　拟　制　造

1. 虚拟制造的定义

虚拟制造提供一种高度集成化的、虚拟的制造环境，其作用是增强生产企业的决策力和控制力。人们把虚拟制造定义为实际生产组织的虚拟模型，而这个实际生产组织在现实生活中可能存在，也可能不存在。虚拟制造不仅包括所有与生产过程、生产过程控制和管理及产品数据有关的信息；还包括部分实际上存在或不存在的、与生产企业有关的信息。虚拟制造借助计算机模型模拟产品的生产制造过程达到辅助产品设计与制造的目的。

提供完整生产制造环境的虚拟制造可分为以下三种模式。

（1）以设计为中心的虚拟制造：为设计师提供能够设计出满足产品设计要求的必备工具。

（2）以生产为中心的虚拟制造：提供研制、分析产品生产制造过程的方法。

（3）以控制为中心的虚拟制造：借助产品生产过程的数字仿真技术，对产品的设计、制造及管理过程进行分析、评估、优化、改进。

2. 虚拟制造的意义

虚拟制造为许多相对孤立的产品制造技术（如 CAD、CAM、CAPP）提供了综合运用的生产制造环境，使不同需求的用户无须自己集成上述制造技术就可以完成产品的部分甚至所有加工制造任务。例如，采用虚拟制造，产品工艺师和制造工程师能够对产品的生产、制造过程进行评估，并将评估结果反馈给异地的产品设计师，由其根据反馈结果对产品进行改进设计。

虚拟制造的另一个重要贡献是促使虚拟企业诞生。虚拟企业曾被定义为了完成某种产品的设计和生产而快速组建的、多学科联合的、有特定目的的小型网络化公司。许多个人和公司通过虚拟制造提供的虚拟环境中的信息共享彼此的技术、资源，对市场中有利的时机进行投资。虚拟制造的主要优点在于为用户营造一个信息非常丰富的环境，加强了产品生命周期各阶段的信息交流。

3. 虚拟制造的应用

（1）钢铁行业。

宝武钢铁引入 AR（增强现实）并打造了"AR 智能运维系统"，为冶金企业带来了全新的设备运维工作方式。该系统结合了 5G、云计算、边缘计算、大数据、人工智能、AR 等新一代信息技术，实现了关联设备的数字信息可视化、精准远程协作与高效过程记录管理。

（2）能源行业。

正泰集团引入 AR 并打造了"AR 配电运维系统"。该系统以可视化方式将作业指导书内容配置导入，现场员工佩戴 AR 眼镜即可扫描设备二维码查看作业指导内容，极大提高了人员现场作业的规范水平及效率。

（3）自动化行业。

中国石化集团东北石油局有限公司在已有 GIS（地理信息系统）的基础上，集成 AR

远程通信与协作插件 HiLeia.PS，打造了"AR 可视化协作管理平台"。该平台主要用于故障诊断和应急维修两个场景，帮助后端专家与现场作业人员进行第一视角的可视化沟通。

（4）重工机械行业。

在海尔上海洗衣机互联工厂，应用数字孪生技术的洗衣机内筒生产线正在运行。通过建立二维数字模型，将现实生活中的场景实景复刻，在虚拟世界中打造一个"双胞胎"，实现生产的改进优化、降低成本、提高效率。

这些案例展示了虚拟制造在提高生产效率、优化生产流程、降低生产成本、提升产品质量等方面的广泛应用和显著效果。随着计算机技术和虚拟现实的不断发展，虚拟制造的应用前景将更加广阔。请注意，以上信息仅供参考，具体应用效果可能因行业、企业、技术等因素而异。

第 21 课　智能制造中的工业大数据决策

在大数据时代，制造业生成的大量大数据具有超高维度的特点。处理这些超高维度的大数据，挖掘其潜在价值，并开发适合智能制造环境的数据流模型是具有挑战性的问题。在"工业 4.0"的背景下，基于大数据分析的方法将为制造业带来更多好处，通过相关新兴技术的相互支持，数据分析过程旨在提高决策透明度。基于大数据分析的决策将根据企业内部结构最大限度地发挥整个制造系统的作用，有效利用制造资源以确保效益最大化。

在智能信息互联和知识驱动的时代，大数据掀起了数字革命的浪潮。基于大数据分析和智能计算的解决方案逐渐用于减少处理大量数据的复杂性问题和认知负担。越来越多的企业采用由大数据驱动的强有力的战略来提高竞争力。大数据驱动技术为如今的制造模式提供了从传统制造向智能制造过渡的绝佳机会。近年来，随着工业工厂的智能发展，大数据分析已成为企业提供工业价值的主要推动力，使工业生产更加智能。从各种来源搜集的数据已应用于工业生产研究。通过数据启用的生产研究已从基于分析模型的研究转向数据驱动的研究。

大数据分析是传统数据分析的一次革命性飞跃。大数据的特点可总结为"五高"：高容量（大量数据）、高速度（数据以高速生成和更新）、高多样性（不同来源生成的数据以不同的形式呈现）、高准确性和高价值（数据中隐藏着巨大的潜在价值）。在大数据时代的制造业中，大数据系统具有实时、动态和自适应的特点。与传统数据分析系统相比，由大数据平台管理的数据来自物理实体世界或虚拟数字世界。由于数据来源具有多样性，因此高效处理数据的能力为制造业提供了更广阔的前景。制造业正在经历信息和智能转型的革命。提供按需通信服务是确保制造系统高可靠性、高可扩展性和高可用性的必要条件。

智能制造不仅涵盖制造领域的许多方面，还整合制造领域和信息技术的各方面，旨在将产品生命周期中获取的数据转化为智能制造，从而对制造的各方面产生积极影响（如智能产品、智能生产、智能服务等）。大数据可以创造实时解决方案，以面对各领域的各种挑战。基于大数据的方法将影响生产系统的产品质量管理，挖掘和分析与产品质量相关的数据可以为制造系统的质量控制及保证提供决策支持。智能制造旨在构建高度集成的协同生产生态系统，可以实时响应整个价值链中不断变化的需求和环境条件。智能制造的核心在于物理世界和数字世界的互联与深度整合。当代制造业的战略重点在于将先进的数字信息技术整合到制造业的各应用领域中。随着"工业 4.0"的到来，制造生产和价值创造过程及业务模式的多个方面都发生了巨大变化。全球制造公司都继续朝着智能方向发展，制定合理、高效的数字战略将稳定提升其竞争优势。

随着数据存储和分析技术的进一步发展，大数据分析是创造制造业主要价值的重要推动力。领导者的决策方法也在不断改变，其主要依靠大数据分析而不是经验来创造更多制造价值。大数据驱动技术为制造业提供了广阔前景，并为未来可持续制造奠定了基础，推动了"工业 4.0"的实践和发展。

在当今竞争激烈的背景下，企业不仅对了解大数据分析的技术方面感兴趣，而且越来越关心充分利用其拥有的数据知识、洞察力及创造力，并有效地将这些知识应用于战略和操作决策及创新过程中。大数据已经成为智能制造领域的研究热点，智能制造关注大数据与协同创新的相互关系，并将大数据作为一种常见的分析视角，以及一个将不同研究流派（开放创新、共创和协同创新）聚合起来的概念。智能制造调查物联网和大数据在业务中管理数字转型方面的作用。智能制造提出大数据和微电网的识别及挑战，并总结微电网的增强领域，如稳定改进、资产管理、可再生能源预测和决策支持。它提供针对数字制造平台集群执行的制造系统相关标准的分析，以确定哪些标准可能与零缺陷制造及其他项目或制造平台设计相关，凸显其可靠性如何用于支持与"工业 4.0"技术相关的不同类型的战略决策。然而，迄今为止，还没有关于在智能制造中智能决策和大数据关系的系统性综述。本书旨在填补这一空白，指出充分利用大数据分析能力对实施智能制造中的成功决策至关重要。

第 22 课　机器人技术与计算机集成制造

1. 智能机器人制造

机器人学是一门研究利用计算机控制机械设备感知和行动的科学。机器人设备种类多样，包括工业机器人、移动机器人、医疗和手术机器人、康复机器人、无人机、服务机器人等。在工业机器人中，机器人操纵器是最常见的设备。工业机器人的发明和应用是基于现代计算机、集成电路和机器人学发展而来的。从通用汽车公司在 1961 年首次使用工业机器人以来，工业机器人的突破时代到来了。如今，机器人在制造业中得到广泛应用，从瑞士 ABB、德国库卡（KUKA）、丹麦优傲（UR）、日本发那科（FANUC）和日本安川电机（YASKAWA）等知名供应商的工业机器人操纵器扩展到包括自动导引车（automated guideol vehicle，AGV）和无人驾驶飞行器（unmanned aerial vehicle，UAV）等机器人。这些机器人的应用包括搬运、喷涂、包装、抛光及新兴的人机协作。因此，机器人已经成功解放了承担工厂中重复和过载任务的人力，将制造业推向了自动化的新时代。然而，机器人最终并没有突破机器人学的最初边界，即通过计算机控制来操作物理世界。机器人的动态、不确定性和灵活性给机器人技术带来了挑战。

对更为先进的机器人技术的渴望催生了智能机器人制造，其中机器人被设计为具有更高智能程度来处理更复杂任务的能力。从物理角度看，机器人由电机驱动，进一步由电力驱动机器人上的各电机。电机产生力并控制机器人的机械结构，遵守机器人动力学。机器人内部的电机旋转，使末端执行器可以根据机器人运动学规则在笛卡儿坐标系中的姿态进行移动。最后，这些移动可以被设计为适用于不同工具的多个制造任务，应用于末端执行器。因此，机器人控制有以下三个抽象级别。

（1）第1级：电机级别。电机控制强调一个电机需要产生的力。通常，电机控制受电机的功率影响，从而影响与末端执行器直接接触的物体。电机的旋转速度也取决于功率。在定制的机器人设备中，终端用户可以通过电机驱动器和微控制器（如通过脉冲宽度调制）为电机功率编程。然而，工业机器人供应商通常限制了电机功率的可用性，为KUKA iiwa、ABB GoFa和UR等有全转矩传感器的协作工业机器人带来了更多可能性。

（2）第2级：运动级别。运动控制，又称运动规划，从AGV和UAV的路径规划到工业机器人的轨迹规划，其在工业机器人应用中扮演着重要角色。顺序姿势代表了在时间维度或笛卡儿坐标系内的运动。机器人的姿势以关节角度时间值的形式体现。关节角度的控制以电机控制为基础。工业机器人通过可编程的运动控制来支持它们的关键功能，如ABB机器人主要使用ABB RAPID编程，KUKA机器人主要使用KUKA KRL作为编程语言。运动级别是电机级别的抽象化，因为它更关注机器人或机器人工具中心点（tool center point，TCP）的运动、路径或轨迹，而电机级别控制瞬时运动。

（3）第3级：任务级别。任务控制将电机控制和运动控制包装成完整的解决方案，用于工业机器人执行任务，如包装、运输、切割、焊接、抛光、喷涂等。一个任务由多个过程组成，对于任务中的机器人，它通常等同于机器人轨迹的计划。一条轨迹是指在笛卡儿坐标系中TCP顺序姿势的曲线，具有起始点和结束点。不同的轨迹，其起始点和结束点会有所不同。通过顺序拼接轨迹，机器人可以在物理世界中生成逻辑动作。然而，轨迹不是一个任务中的唯一元素。触发输入/输出（I/O）信号、在用户界面上发出警告或写入数据库也是任务的一部分，尤其是在针对全面工作单元的情况下。总体而言，任务级别将电机级别和运动级别抽象化，因为它受到关节角度和轨迹上的转矩控制。

电机级别、运动级别和任务级别的控制容易混淆。事实上，它们需要不同领域的不同学科的不同主题作为理论依据。这些学科包括但不限于电机工程、自动控制、生产计划、机器人运动学、机器人动力学、规划算法和优化。由于工业机器人的商业机密和专利问题，因此通常会隐藏电机控制，强调工业用户的运动控制和任务控制。然而，在静态环境下控制运动和任务不能称为智能。面向新一代智能机器人制造，机器人需要适应动态和多变的环境，人类的干扰使计划的轨迹不再适用于机器人跟随。因此，机器人需要灵活地与人类互动，如在接近终点时调整运动。这种灵活性不仅与运动级别有关，还与任务级别有关，当人类改变计划的过程（如过程的顺序或时间消耗）时，挑战在于干扰带来了不确定性，导致一个更棘手的问题——机器人需要具备高级智能来覆盖这些随机属性。机器学习正是一种有望提供强大表示力的信息技术，可以提升智能级别。

2．机器学习

机器学习是一个将计算机科学与统计学结合，并与信息理论、信号处理、算法、控制理论和优化理论相联系的领域。如今，机器学习已经成为人工智能中具有吸引力的领域之一。机器学习的范式非常简单，即用数据拟合函数，并将拟合的函数用于对新数据的近似。从一个弱函数到一个强函数的发展，利用给定数据的过程称为学习，这类似于动物的学习过程。由于这种发展通常由计算机等机器完成，因此相应地提出了"机器学习"。在机器学习中，拟合函数以原始数据或手动制作的特征作为输入，并通过自身传递信息生成输出。在深度学习出现之前，特征工程是通过手动或使用预定义规则从数据中提取特征的。这个过程已被更复杂的模型（如神经网络）取代，具有更强的表示能力。然而，神经网络不是这种函数使用的唯一模型，线性模型、二次模型、概率分布的表达式（如高斯估计或高斯混合模型）甚至离散表格也很常用。对于神经网络来说，卷积神经网络（convolutional neural networks，CNN）、循环神经网络（recurrent neural network，RNN）或图神经网络（graph neural network，GNN）是深度神经网络（dense neural network，DNN）之外的选择。不同模型的提出源于问题本身，如果一个问题容易建模或可以使用已知概率分布有效处理，那么可以选择这些模型，然后学习参数。对于棘手的问题，神经网络更受欢迎，因为有证据表明，从理论上讲，神经网络可以表示任何非线性函数。模型的可用性通常取决于数据的维度。例如，从图像中提取语义信息需要具有更强表示能力的逼近器，如深度神经网络，因为图像通常具有比常规数值更大的数据输入量。至于神经网络，它也因不同的任务而异。CNN专为以几何形状组织的数据（如图像）设计，而RNN适用于顺序数据（如文本和声音）。无论选择哪种模型作为机器学习的函数，学习阶段都是根据给定的数据来估计模型中的参数的。这些参数［如模型类型和内部设置（高斯分布的组合、神经网络中的层数）］称为超参数。

输入数据可以是标量、矢量、矩阵或张量，如数值数组、文本、音频或图片。输出也可以根据用户的要求具有不同的维度，有时甚至与输入相同，如变分自动编码器（variational auto-encoder，VAE）和生成对抗网络（generative adversarial network，GAN）。事实上，机器学习用其函数将数据从输入映射到输出。不同的映射会产生不同的应用。例如，将图像映射到标签是识别，将一种语言映射到另一种语言是机器翻译。将任何数据从输入映射到输出都可以用于机器生成或维度降低。

找到函数参数正是机器学习的方法。然而，参数调整与机器学习技术的类型密切相关。机器学习方法通常分为三类：无监督学习、监督学习和强化学习。无监督学习侧重于没有监督信号的信息处理，如聚类。监督学习提供可以用于计算损失函数的监督信号，函数的参数是在给定成本函数的优化的情况下调整的；梯度方法是广泛使用的优化方法，因为它可以有效地由计算机计算。强化学习基于试错范式。以人类学习为例，监督学习就像从书本中学习，而强化学习就像从经验中获取知识。不可否认的是，不是所有知识都可以从书本中学到的，强化学习可以说是机器人学习的一个关键方法。

3. 机器人学习

未来制造业中的机器人将不再只被安置在固定的工作站中执行成千上万次相同的程序，而将涉及变化的环境，而且无法完全对其执行的任务进行预编程。这个挑战要求机器人处理不确定性问题并采取相应的行动。机器人学习是指通过人类的帮助使机器人自己学习，利用传感器和电机扩展来增强其初始智能，以应对新的环境。

机器人学习是在机器人领域多种机器学习技术的综合，强调使用机器学习技术进行机器人学习和行动。与常见的机器学习不同，机器人学习强调以行动作为输出，同时观察环境作为输入。深度学习帮助机器人处理非结构化环境，而强化学习提供了机器人行为的形式化描述。这种范式类似于人类的行为方式,人类通过感官器官感知环境,通过神经系统与大脑相连。例如,环境中的光线由视网膜上的光敏细胞感知,然后通过生物电信号传递到大脑皮层,大脑皮层向相应的肌肉发送信号以产生运动,这个过程形成了一个封闭式循环,就像自动控制一样。再如,当一个人要抓住一杯牛奶时,先用眼睛定位杯子,再在移动手臂时看向手。机器人设备在封闭式回路中也表现出类似的行为。机器人利用摄像头观察环境,并使用计算机处理感知数据,然后通过算法向机器人控制器发送信号来驱动机器人操纵器。然而,没有哪种动物天生具备这种技能,都是在成长过程中学会观察并做出行动。此外,这种学习方式显然不是完全监督的,更像是从经验中学习或试错学习的,强调了强化学习。机器人学习受到动物学习和适应环境的启发。

虽然一般的机器学习侧重预测和分类,但是机器人学习强调机器人系统中配置值的输出,如时间关节值或任务导向的行为过程。这些信号输出生成机器人行为的运动属于机器的运动控制范畴。机器人学习通常将来自物理世界的实时传感器数据作为输入,即所谓的观察。策略或控制器将观察映射到由机器人配置指示的动作中,机器人可以操纵物理环境,并导致环境的状态转换,然后通过迭代传递给传感器生成新的观察。

马尔可夫决策过程(Markov decision process,MDP)是一种用于决策和状态转换的简化数学模型。由于决策是一个接一个地作出的,因此又称顺序决策过程。具体来说,与由直接损失函数监督的机器学习不同,机器人学习依赖环境的累积奖励(或成本)信号。当机器人学习时,作决策的机器人通常称为代理,代理接收观察并运行策略。根据感知条件的不同,观察可以分为全观察和部分观察。策略涉及采样方式。环境模型的可用性区分了基于模型的机器人学习和基于无模型的机器人学习。

机器人学习出现在不同的机器人技能中。操纵学习侧重于通过持续学习操纵各种物体。目前,机器人擅长在重复的刚性几何形状中拾取和操纵物体,但在运动变化方面较弱。有目的的推理是机器人学习的另一个挑战,特别是在与人类互动时,机器人注重推理人类活动,以增强人机协作。采用基于数据驱动的机器人学习(如神经网络)在机器人抓取、电机控制、运动规划、通过触感识别等方面表现出出色的能力。机器人模仿学习也是一个经典且受欢迎的研究课题,如从人类教师表现或其他不适合模仿的任务中学习。这些机器人技能还在制造领域中显示出巨大的潜力,使制造业从大规模生产转向大规模定制。在智能机器人制造中,机器人学习已经在电机级别、运动级别和任务级别中展现出强大的能力。

第 23 课　产品测试与质量控制

1. 产品测试

产品测试不但是所有工艺控制过程中至关重要的部分,而且是整个工艺过程监控中的质量检验部分。如果产品质量控制计划制订合理、执行情况良好,那么其性能有所保障,无须对其进行总体性能测试。因此,产品质量的总体测试只是用户取货之前必须履行的合同要求。但产品测试不仅是一个验证过程,从整体上说还是所有部分检验的总和。通过产品检验能搜集数据,这些数据可支持、验证产品设计理论;能为以后产品的改进设计提供依据，使产品的性能更加优越；能对产品的更新换代进行性能和成本评估。此外，在产品的实际设计过程中，不是所有设计参数都能被计算和预测，所以产品测试也是检查设计方法合理性的一种方法。

产品测试工程师应与产品设计工程师密切合作，为产品的测试提供有用的数据，同时必须与各工序的制造工程师协同工作。在产品的测试过程中，经常会发现设计工作中的一些错误，这些错误是产品制造过程中需要重点修改的地方。因此，制造工程师和设计工程师都很关心产品的测试结果。

对于复杂的产品来说，产品测试已经成为产品整个工艺控制过程中的重要组成部分。它为公司提供了较高的可信度，使产品具有用户希望的性能。由于产品测试能帮助公司在用户中建立良好的声誉，因此它是一个极有价值的市场调控工具。

2．几何误差

几何误差是指单个机械零件的误差，如线性轴承运动的直线度。与几何误差相关的是准静态时的表面精度，其与零件表面的相对运动有关。零件的几何误差可以减小，也可以保持不变（系统误差）；可能滞后出现（如齿轮隙），也可能随机产生。影响零件几何误差的因素有很多，如表面直线度（图 23.1）、表面粗糙度、轴承的预加载荷、弹性运动设计方法和结构设计原理等。

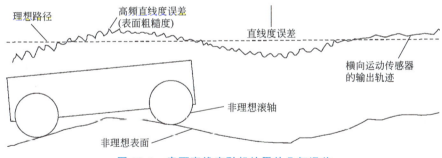

图 23.1　表面直线度引起的零件几何误差

3．质量规划

质量规划是工艺控制中的规划和决策行为，是指工艺规划或检测规划。工程师提出充分的执行计划是为了保证产品的性能达到设计要求。质量规划工程师以产品的实际生产效率和检测结果为指导原则，确定在制造过程中进行检测和非破坏性测试的环节以及检查和测试的内容，并且依据产品的设计要求判断产品是否合格。

通常产品的制造与设计会存在一定的偏差。有些偏差对产品的质量影响较大，有些则没有较大影响。质量规划工程师的主要职责是对产品制造和设计过程中的偏差进行评估，并对影响产品质量较大的偏差找出恰当的修改方法。作为产品设计与制造偏差的评判者，质量规划部门可以通过数据库对产品的加工工序及制造情况进行评估，因此稳定的产品质量是能够达到的。通过这种方法，质量规划部门可以向管理部门汇报产品的质量状况。

利用统计方法对上述偏差进行评估，可以了解产品的质量状况，也可以估算改进这些偏差需要的成本。维修作为产品制造耗材的组成部分，也是产品质量水平控制的重要度量方式。产品制造过程中的浪费状况是对操作工及前面工序的操作员工作态度的衡量指标。

由于浪费多说明管理差，因此质量规划工程师的责任是减少产品制造过程中的浪费，做好预算并制订检测方案。

4. 质量控制

从传统意义上讲，质量控制是制造与设计的联系纽带。它的功能是对产品的制造进行规范设计，同时制订用于产品制造、操作的质量规划。此外，质量控制负责向管理部门建议允许损耗的范围。这个范围是根据产品设计的复杂程度，尤其是误差范围内精度的级别确定的。传统的质量控制方法通过设置不允许超出的负预算来控制产品的制造损耗，并以此建立产品检测和维修的准则。

为了保证产品质量，建立了与市场和用户有关的质量控制文件系统。这一做法赋予了产品质量控制新的使命，与传统意义上的产品质量控制方法是有区别的。

质量控制的目的是借助产品各阶段的制造规范，确保制造出符合要求的产品。虽然质量控制与制造过程直接相关，但它通常还与市场调控功能中的用户职责有关。许多企业已经有选择地建立独立的产品质量控制规范，明确相应的质量控制技术职责，即在企业内部实行独立的产品质量控制体系。

第 24 课　机电一体化

机电一体化最初是在 20 世纪 70 年代由两个工程类学科——机械学和电工电子学结合而成的，但是随着控制工程和通信领域的快速发展，"机电一体化"一词已经被用于描述机械工程学、控制工程学和电子学三大学科的集成，其主要目标是研究由电子学控制的机械产品的制造，同时被认为是机械工程学、电子学及智能计算机控制学在工业产品设计制造及加工过程上的集成。随后，由于这些工业产品具有智能和灵活的特征，因此人们经常将这些工业产品称为智能化或灵活型的产品。从而机电一体化可定义为在工业产品的设计和制造加工过程中应用电子学技术、智能计算机控制技术与机械工程学等的集成学科。

在工程上，机电一体化是由机械学、电子学、计算机科学和制造工程学等学科产生的一个交叉学科。"机电一体化"课程已经在工程教学中作为一门必修课程，机电一体化领域已经作为新的工业产品开发基础日益被人们接受。机电一体化技术主要包括机电一体化的系统建模、仿真、传感和评估系统、驱动和执行系统、系统行为分析、控制系统及微处理系统。

1. 机电一体化系统

机电一体化系统处理工程、科学和技术上出现的还没有被解决的基准程序技术和新兴的问题。机电一体化系统使用高性能的微处理器、数字信号处理器（digital signal processor，DSP）、驱动电子元件、集成电路（integrated circuit，IC）来进行系统设计、优化、建模、仿真、分析和建立有效模型等工作，它是一种可以获得智能化、高性能机电系统（包括机械和加工），可以实现智能化和运动控制的综合性研究。它具有多学科交叉的特点，集成了电气工程、机械工程和计算机工程等学科，如图 24.1 所示。

在机电一体化系统设计中，具有重要意义的问题是系统结构的开发设计，包括硬件（如执行器、传感器、装备、电子产品、集成电路、微控制器和 DSP 等）和软件（如实现

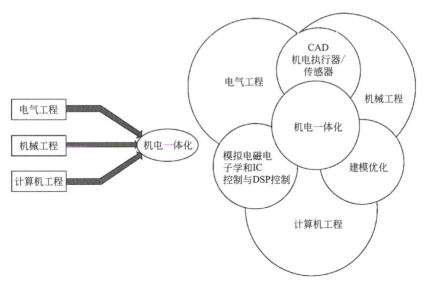

图 24.1 机电一体化系统

传感与控制、信息流与数据采集、模拟仿真、可视化处理和建立有效模型等的软件环境和计算机算法）的选择。通过分析复杂的模型和高级的生物系统的范例来设计具有艺术特性的人造的机电一体化系统，并保证这一集成设计继续深入进行。

现在的工程设计趋于高级的机械系统的集成分析、设计和控制。机电一体化系统的范围已经得到扩展，除执行器、传感器、驱动电子元件、IC、微处理器、DSP 外，还集成了输入/输出设备及其他许多子系统。

机电一体化系统的最终目标如下。

（1）保证最终结果的一致性及具有学科交叉的特征。

（2）拓宽传统的机械、电机系统、驱动电子元件、IC 和控制理论，以满足高级的硬件和软件需求。

（3）获取并拓展多学科的工程核心集成领域。

（4）将电机系统、驱动电子元件、IC、DSP、控制理论、信号处理、微型机电系统和纳米电子系统技术融合，以构成机电一体化系统的综合体系。

高性能电机系统被认为是机电一体化系统的重要基础。执行器和传感器（如电机驱动装置）、驱动电子元件和 IC、微处理器和 DSP、高级硬件和软件系统的统一分析基本没有被引入工程课程教中。机电一体化系统在传统的微型和纳米规模的电机系统中是一个突破性的概念，已经被用于破译、集成和解决大量工程问题。

2. 机电一体化系统分类

机电一体化系统可分为传统的机电一体化系统、微机电系统（MEMS）、纳米机电系统（NEMS）。传统的机电一体化系统与 MEMS 的工作原理和基础理论相同，而 NEMS 研究不同的概念和理论。特别值得一提的是，工程设计人员应采用经典力学和电磁学来研究传统的机电一体化系统和 MEMS。定量分析和纳米电机学已应用于 NEMS 中。一些基本理论可以用于研究传统的、微观的和纳米的机电一体化系统中的效果、过程、现象，如图 24.2 所示。

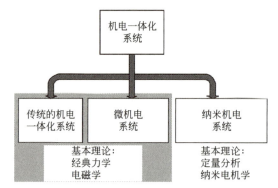

图 24.2 机电一体化系统的分类及其所运用的基本理论

第 25 课 微机械技术导论

1. 微机械技术的发展

微机械技术在 20 世纪 80 年代初只是科学家提出的一种设想,然而微制造技术(如半导体微细加工技术)及其他相关技术(如设计、材料、测量、控制、传感技术、信息处理、计算机、能源及系统集成等)发展到一定程度后,美国加州大学伯克利分校和麻省理工学院的研究小组在 20 世纪 80 年代后期利用半导体制造技术成功研制出直径约为 $100\mu m$ 的静电微型电动机。这一成果在国际上掀起了微机械技术研究的热潮,受到了各国专家的极大重视,因此微机械技术成为 20 世纪出现的一项高新技术。

人们预测微机械技术将会给工业领域的发展带来一场大的变革,许多发达国家和地区都把它看作对繁荣经济和国防安全至关重要的技术,成为优先支持的项目,投入巨额资金进行研究开发。这促使微机械技术得到迅猛发展,也取得了一定的成绩。例如,斯坦福大学成功研制出直径为 $20\mu m$、长度为 $150\mu m$ 的铰链连杆机构,$210\mu m \times 100\mu m$ 的滑块机构,转子直径为 $200\mu m$ 的静电电动机及流量为 $20mL/min$ 的液体泵。东京大学研制了尺寸为 1cm 的微型爬坡机构。名古屋大学成功研制出一种不需要电缆、用于微小管道检测的爬行机器人,它是通过管外的电磁线圈产生的磁场来控制其运动的。在我国,微机械技术也引起了广大学者的普遍重视。目前研究集中在机械零件、集成传感器、光学元件、驱动器等方面并取得一定的成效,现在正进行相关应用方面的研究。上海交通大学的科学家基于他们构建的人体器官芯片(OOC)模型,开发了一种磁控微纳米机器人的多阶段输送策略。他们借助微纳 3D 打印技术,构建了分层血管化 OOC 模型,可以再现人体不同器官的结构和功能。该团队可以将磁导微纳米机器人注入 OOC 中,并利用外部磁场在血管内导航它们。这些体外模型的应用包括促进药物筛选、降低药物研究成本及提高药物开发效率。它们还可以用于医疗,如为肿瘤患者建立体外 OOC,使其作为患者的替代品来尝试不同的药物并指导临床用药。为解决微型零件加工需求,速科德研发了超高精密设备 KASITE-SKD 系列微纳加工中心,加工余量范围为 $20nm\sim100\mu m$。

2. 微机械技术的基本特征

机械微型化带来了尺度效应的问题。随着机构的尺寸减小,其物理量并非成比例地缩小；当尺寸缩小到一定程度时，宏观机械常使用的计算方法和理论将不再适用。所以，微机械技术具有以下基本特征。

（1）表面力起主导作用。体积力（如重力、电磁力）与特征尺寸的高次幂成正比；表面力与特征尺寸的低次幂成比例，如摩擦力、表面力、静电力。微机械体积小、质量轻，其表面力与体积力的比值较大，因此与体积力相比，表面力成为主要载荷。同样，与重力相比，静电力成为主要载荷，这称为机械微型化的尺度效应。所以，在微机械中经常使用静电力驱动。此外，与重力相比，摩擦力对微机械的影响程度比宏观机械大得多。

（2）微机械并不是宏观机械原型的缩小。传统机械各部件的复杂程度一般各不相同。因此，用几何方式缩小这样的机械结构对复杂部件很难做到，对高智能自动设备更是如此。设计微机械不必追求复杂的机械结构，而应着眼于用多个结构简单的机械元件（包括带有传感器和人工智能的器件）来完成复杂的工作。

（3）不同的能量供给方式。对于具有移动和转动功能的微机械来说，电缆常常会限制其运动范围，所以一般不采用电缆供电。目前微机械一般采用静电力提供能量，有时也使用振动激励的方式提供能量（如利用压电特性、电磁特性及形状记忆合金驱动）。

因此，微机械的研制需要一些新理论和新方法。例如，微机械在运行过程中受到的阻力形式发生变化，需要一些新的机构设计原理和控制方式；运动及动力的形式发生变化，需要一些新的驱动方式；器件的结构微型化，需要一些新的制造技术和加工方法；等等。

3. 微机械技术的应用

微机械技术的兴起和发展表明它有较广泛的应用前景，主要应用领域如下。

（1）在机械设计领域，微机械技术用于设计微机械系统中的各种微型机械结构，如微齿轮、微连杆机构、微滑块机构等。

（2）在仪器仪表领域，微机械技术用于设计压力传感器、加速度传感器等。

（3）在流体控制领域，微机械技术用于设计微泵、智能泵等。

（4）在微光学领域，微机械技术用于设计光纤、光扫描仪和干涉仪等。

（5）在超大规模集成电路制造领域，微机械技术用于制造真空机械手、微定位系统和气体精密控制系统等。

（6）在信息机械领域，微机械技术用于设计磁头、打字机头和扫描器等。

（7）在机器人领域，微机械技术用于设计微型管道检测与修复等极限工作环境的微机器人和多自由度机械手等。

第 26 课　工业机器人

工业机器人是一种可自动控制、可编程、多功能的、由多个可重复编程的坐标系操纵的机器装置，以固定或移动的方式应用于工业自动化中。

工业机器人的主要优点在于可重复编程和具有多功能性，因为大多数功能单一的工业机器人不能满足这两种要求。可重复编程包含两层含义：工业机器人根据设定的程序工作，并且这个程序可以被重写，以适应多种任务。多功能意味着工业机器人可以具有多种功能，以适应通用的不同程序和不同工具。

经过多年发展，工业机器人已经进入工厂来完成许多单调且不安全的操作任务。因为工业机器人可以比人更快、更准确地完成某些基本任务，所以其广泛应用于制造业。

1. 工业机器人的结构

工业机器人的结构主要包含四个部分：操纵器、末端执行器、动力供给系统和控制系统，如图26.1所示。

操纵器是机械系统，进行类似于人的手臂的运动。它通常有肩关节、肘关节和腕关节。它能旋转或平移，以一定的灵活性在各个方向上伸缩。

工业机器人按操纵器的基本机械构造可分为笛卡儿坐标机器人、圆柱坐标机器人、球坐标机器人和铰接机器人。笛卡儿坐标机器人可以在立方体或矩形范围内把手爪移到各位置，这可界定为它的工作范围。圆柱坐标机器人可以在定义的圆柱体范围内移动手爪，可以通过在 X 轴和 Y 轴方向的线性运动和相对于 Z 轴的

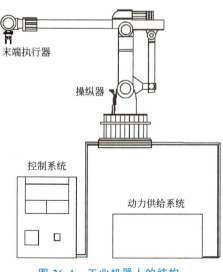

图26.1 工业机器人的结构

一定角度的旋转来界定工作范围。球坐标机器人通过两个旋转运动和一个线性运动定位腕关节。铰接机器人具有一种不规律的工作范围，这种工业机器人具有两个重要变量：垂直铰接和水平铰接。

末端执行器与腕关节相连，也称手臂夹持器，是类似于手的操作装置。末端执行器一般根据特殊需要设计。机械手是最常用的，一般装有两个或两个以上手指。末端执行器的选择取决于有效载荷、环境可靠性和价格等因素。

动力供给系统是移动机械手，控制关节，操作末端执行器的驱动器。动力供给方式有电力、气动、液压三种。每种动力供给方式都具有自身的特性、优点和局限性。根据动力供给系统的设计和用途选择交流电动机或直流电动机。电动机把电能转化为机械能，为机器人提供能量。大多数新型工业机器人采用电力驱动。气动应用于高速、非伺服工业机器人，也应用于驱动工具，如手爪。液压驱动用于较重的提升系统，尤其是精度要求不高的场合。

控制系统是通信和信息处理系统，它发出指令驱动工业机器人动作。它是工业机器人的"大脑"，向动力供给系统发出信号，把工业机器人的手臂移到特定位置，并驱动末端执行器。它也是工业机器人的"神经系统"，对工业机器人的所有运动及动作发送的指令序列是可重复编程的。

开环控制系统是控制系统的最简单形式，它通过预定的指令逐步控制工业机器人。这样的系统不具有自我纠错能力。而闭环控制系统由反馈传感器产生信号，信号反映被控目标的当前状态。通过对比反馈信号与程序设定值，闭环控制系统能引导工业机器人向准确的位置运动并实现期望的状态。闭环控制系统可以使被控目标与预定值的误差最小，末端执行器可以精确地运作。

2. 工业机器人的分类

工业机器人的尺寸、形状、坐标数、自由度和设计构造多种多样。每个因素都影响工

业机器人的工作范围及它能够运动和执行指定任务的空间区域。广义的工业机器人分类如下。

（1）固定顺序机器人和可变顺序机器人。固定顺序机器人（也称拾取和定位机器人）是为完成一系列特定的操作设计的。它的运动是点到点的，并且可以循环。可变顺序机器人是为完成特定顺序设计的，也可为其他操作重新编程。

（2）示教机器人。操作者可以按期望路径引导示教机器人及其末端执行器运动。示教机器人可以记忆和记录运动的顺序及路径，并在没有操作者引导和示范的情况下重复做这些动作。

（3）数字控制机器人。数字控制机器人的编程和操纵很像数控机床，这种机器人由数字数据伺服控制，运动顺序易改变。

（4）智能机器人。智能机器人能够执行一些人类才能完成的任务。它可以配备各种传感器以具备视觉和触觉功能。

3. 工业机器人的应用

由于工业机器人是一种很特别的生产工具，因此其应用十分广泛。这些应用可以被划分为三类：材料处理、材料搬运和自动装配。

在材料处理中，工业机器人用工具来加工和处理原材料。例如，工业机器人的工具有钻头，可以在原材料上钻孔。

材料搬运包括装载、卸载和转移制造设备上的加工零件。这些操作可以由工业机器人可靠地重复执行，因此提高了加工质量，减少了废料损失。

自动装配系统能合并自动测试、工业机器人自动控制和机械处理，以减少劳动成本，提高产量，消除人工操作的危险性。图26.2所示为应用于自动装配的SCARA工业机器人。

图 26.2 应用于自动装配的 SCARA 工业机器人

第 27 课 小小机器人兵团

一群恐怖分子强行闯入一栋办公大楼，劫持数名人质，他们封锁各出入口，并遮住窗户，外面的人完全看不到他们有多少人、携带什么样的武器、将人质关在哪里。但突然间，特警队闯了进去，将来不及拿起武器的恐怖分子一网打尽。自信与果断的行动需要确切的信息，特警队是如何得到这些信息的呢？

答案就是一组合作无间的小型机器人，它们经由通风系统偷偷溜进大楼，从输送管井然有序地到处移动。它们有些配备监听对话的麦克风，有些配备小型摄影机，还有些配备传感器，可嗅到空气中的化学或生物药剂。它们采用合作的方式，将这类实时信息以无线电传回有关当局。

挑战就是要开发出微小的侦察机器人，以便士兵随身携带，像爆米花一样撒在适当的地点。消防队员和搜救人员可以留在大后方，将这些机器人丢到窗内，让它们到处跑，寻找受困的伤者或嗅出有毒物质。

1. 蚂蚁雄兵

原则上，比起笨重的大型机器人，小型机器人具有许多优势。它们可以爬进管道探查倒塌的大楼，也可以藏在不明显的隙缝里。一群组织良好的机器人可以相互交换传感器信息，相互支持，共同攀越障碍物，或是在跌倒后爬起来。团队指挥官可以评估状况，决定送出机器人的数量。如果一个机器人有故障，也不会因此拖垮整个群体，其余的成员还可以继续行动。

但是，小型机器人需要一种全新的设计观。不像体积较大的机器人，小型机器人没有充裕的电源和空间，也无法容纳执行特定任务所需的全部组件，即使只是携带摄影机等小物件也可能被压垮。因此，传感器、处理能力及机械强度要分散到多个小型机器人身上，而它们必须通力合作。这类机器人就像群体里的蚂蚁，单独的时候很脆弱，但联合起来极具战斗力。

2. 定位团队

小型机器人有一件非合作不可的事就是定位，即算出团队的位置。大型机器人有多种定位方法，如全球定位系统（global positioning system，GPS）接收器、固定信标及视觉地标辨识。此外，它们的处理能力足以比对目前传感器的信息与原有地图。

定位方法虽多，但没有一样能确实用在小型机器人身上。毫米机器人的声呐范围有限，只能测出约 2m 的距离。它的体形太小，携带不动 GPS。至于推算航法（利用测量车轮速度记录位置的方法），也会因其质量太轻而行不通。即使是地毯编织方向这种微不足道的小事，也可能大幅度影响它们的移动，使里程读数不精确，就像汽车里程计在结冰的湖面上无法显示精确的距离一样。

因此，必须想出新的技术，我们开发出来的是一种小型化 GPS。这种技术不使用卫星定位，而是利用音波测量同组机器人的距离。利用交替传输与听取的方式，机器人计算出彼此的距离。每次测量需要大约 30ms。团队的指挥（可能是基地或是大型机器人，也可能是部署毫米机器人的母舰机器人）搜集所有信息，并使用三边测量法计算机器人的位置。这种定位方法的优点在于毫米机器人不需要固定的导航参考点，它们可以进入一个不熟悉的空间，然后自行勘测。在绘图过程中，几只特定的毫米机器人会把自己当作信标，暂时保持不动，而其他几只在附近走动，绘出地图并避开物体，同时测量自己相对于信标的位置。彻底探勘信标周围的区域之后，它们的角色互换，原来探勘的机器人会移到适当位置变成信标，原来的信标组开始探勘。这种方法很像儿童玩的"跳马背"游戏，而且无须人类介入就能执行。

3. 联结战车

路上的障碍是小型机器人必须合作的另一个因素。由于体积的关系，小型机器人容易受制于随处可见的杂物，它必须应对石块、污泥或碎纸片。标准的毫米机器人可跨越高度约为 15mm，所以，一支铅笔或小树枝就可能挡住它的去路。为了避开这些限制，我们设计出一种新型毫米机器人，它们可以像火车车厢一样联结。这种新型毫米机器人，每只约 11cm 长、6cm 宽，看起来就像是缩小版的战车。它们通常独立行进，能够灵活绕过小型障碍物，但需要跨过沟渠或爬楼梯时，它们可以联结起来，形成一串锁链。

这串锁链的灵活功能来自毫米机器人之间的联结接头。毫米机器人的联结接头可不像火车车厢联结或是汽车后面的拖车挂钩，它有一个强力的电动机，可以上下转动联结接头，转矩的力量足以举起几只毫米机器人。要爬楼梯时，这串毫米机器人先用力挤向楼梯的底部，比较靠中间的几只毫米机器人会以悬臂方式举起前面的机器人，而到达顶端的毫米机器人可以把下面几只拉上来。目前，这个程序还必须由人类遥控，但假以时日，这一串毫米机器人应该能自动爬楼梯。

研究人员原本比较注重硬件的开发，但现已开始转向控制系统设计的改良。研究的重点从控制少许个体转移到管理几百或几千只个体，这是一种截然不同的挑战，需要相关领域的专门技术，如经济学、军事后勤学甚至政治学。

我们设想的大规模控制方法之一是透过阶层式的管理，采用类似军队管理的方式，将机器人分成几个较小的团队，各团队有控制指挥的队长，队长上面还有层级更高的指挥官。毫米机器人已经可以接收大型机器人的指示，这些长得像坦克车的大型机器人的奔腾（Pentium）处理器足以完成绘图与定位的复杂计算。这种比较大的机器人可以拖着一串毫米机器人，就像母鸭带小鸭一样，必要时将毫米机器人部署到适当的区域。管理大型机器人还有更大的全地形车辆机器人，这些全地形车辆机器人拥有数台计算机、摄影机和GPS。这个想法就是让大型机器人部署小型机器人，进入大型机器人进不去的地方，然后留在附近提供支持与指示。

小型机器人的发展还有很长一段路要走，还没有任何小型机器人团队能走出实验室，在大楼的走廊漫游，四处寻找危险物品。虽然这类机器人有很大的潜力，但是它们目前功能有限，只能算是新奇的小玩意儿。随着科技从军事应用慢慢扩大到其他用途，我们可以预见，小型机器人的能力将会大幅度提升。由于团队合作，它们可以具有各种功能，模块化的设计也便于针对特殊任务修改。无论怎么说，与它们一起工作都很有趣。

第28课　汽车发动机导论

传统燃油汽车由发动机、底盘、车身和电气设备组成。发动机用来为汽车提供动力，又称汽车的"心脏"。

1. 发动机分类

根据设计特点的不同，发动机可以按以下方式分类。

(1) 气缸数量。

根据气缸数量的不同，发动机可分为三缸发动机、四缸发动机、六缸发动机、八缸发动机、十二缸发动机。气缸数量是决定发动机功率和燃油效率的因素之一。常用的是四缸发动机和六缸发动机。三缸发动机能够满足节能减排要求，受到汽车制造商的关注。

(2) 气缸排列方式。

发动机气缸排列方式有直列式、水平对置式、V形。也使用更复杂的排列方式，如W8、W12、W16、W18。直列式发动机采用直列式气缸，适用于三缸发动机和四缸发动机。随着气缸增加到六个，V形发动机适用于减小发动机的长度。当气缸进一步增加时，出现了 W 形发动机，它是一种新开发的发动机，结构复杂，制造成本高。

（3）燃料类型。

燃料主要有汽油、柴油、生物柴油和乙醇等。汽油和柴油是常用的汽车燃料，在世界各地都被用来为汽车提供动力。使用植物油或动物脂肪生产的柴油称为生物柴油。乙醇虽然没有被广泛用作一般的汽车燃料，但是它作为添加剂加入汽油。因为乙醇是一种从玉米和甘蔗等可再生资源中提取的低成本燃料，所以许多汽车制造商设计可以用乙醇作为燃料的汽车。随着传统能源价格的上涨，世界越来越重视环境保护，政府、消费者和汽车企业已经转向低排放汽车的生产。

2. 汽油发动机

汽油发动机包括两大机构和五大系统，两大机构分别是曲柄连杆机构和配气机构，五大系统分别是起动系统、电控燃油喷射系统、点火系统、冷却系统和润滑系统。

（1）曲柄连杆机构。

曲柄连杆机构主要包括气缸盖罩、气缸盖、气缸体、气缸套、气缸衬垫、油底壳、活塞、连杆和曲轴等部件，其作用是把活塞的往复运动转变为曲轴的旋转运动来驱动汽车。

活塞是由铝合金制成的圆柱形空心零件，其在气缸内的往复运动将膨胀气体的能量转化为机械能。连杆连接活塞和曲轴。曲轴是发动机中的重要零件之一，它被固定在气缸体的曲轴箱中。飞轮是位于曲轴一端的旋转盘，通过其惯性减少做功冲程引起的振动，并在起动机与之啮合时起动发动机。

（2）配气机构。

配气机构的作用是定时开关气门，气门的开关是由凸轮轴的转动实现的。气门开关的时间及持续开启的时间称为配气正时，以曲轴转角表示。因为当发动机转速不同时，对配气定时的要求是不同的，所以为了使高速和低速都能实现最佳配气正时，在汽车发动机上开发出可变配气正时。

大部分发动机有两根凸轮轴，一根是进气凸轮轴，另一根是排气凸轮轴。进气凸轮轴用于实现进气门的开关，而排气凸轮轴用于实现排气门的开关。另外，当两个凸轮轴布置在气缸盖上时，形成汽车制造商制造的主流双顶置凸轮轴。为了提高进排气效率，一般每个气缸上都有两个进气门和两个排气门。此外，在一小段时间内，排气门保持开启时，进气门也保持开启，即排气冲程的结束和进气冲程的开始在短时间内重叠，这种现象称为气门重叠。

（3）起动系统。

起动系统主要由起动机、电磁开关、控制电路、起动机继电器、点火开关（起动开关）、蓄电池和起动机电路组成。其中，蓄电池也称铅酸蓄电池，是一种产生电压和传递电流的电化学装置。首先，将钥匙插入点火开关，转到起动位置，少量电流通过空挡安全开关流向起动机继电器或起动机电磁线圈，使大电流通过蓄电池流向起动机电机。然后，起动发动机，使活塞向下运动，产生吸力将混合气吸入气缸，点火系统产生的电火花点燃混合气。如果发动机中的压缩气体压力足够高，并且上述都在正确的时间发生，发动机就会起动。最后，一旦发动机起动，交流发电机就为汽车的所有电气设备供电。

（4）电控燃油喷射系统。

电控燃油喷射系统主要由燃油供给系统、空气供给系统和电控系统三部分组成。

燃油供给系统主要由油箱、电动燃油泵、燃油滤清器、油压调节器、燃油分配管和喷油器等组成。因为油压调节器能够使喷油压力保持恒定，所以喷油器的喷油量主要取决于喷油脉宽，而喷油脉宽是由各种传感器将信号传给电子控制单元（ECU）来实时调整的。

空气供给系统的任务是向燃油发动机提供与发动机负荷适应的新鲜空气，以便与从喷油器喷入进气管或气缸中的燃油形成优质的可燃混合气。空气供给系统主要由空气滤清器、空气流量计、节气门、进气歧管等组成。驾驶人通过踩下加速踏板来控制节气门的开度，以决定进入气缸的空气量。

电控系统主要包括各种传感器、ECU和喷油器。ECU通过监测发动机传感器来精确确定喷油器的喷油量。

（5）点火系统。

点火系统主要用于在准确的时间向发动机提供强烈的电火花来点燃燃烧室的混合气。点火系统一般分为无分电器的点火系统和有分电器的点火系统。其中，无分电器的点火系统始终由计算机控制，不包含任何运动部件，大大提高了点火系统的可靠性。无分电器的点火系统的每个气缸都安装一个点火线圈，或者每两个气缸共用一个点火线圈。

独立点火系统的每个火花塞都配备单独的点火线圈，直接安装在火花塞上，以使火花尽可能高，提高了点火系统的可靠性。由于性能、排放量和维护等因素，独立点火系统已成为热门设置。将单独的点火线圈直接安装在每个火花塞上消除了对长且笨重的高压火花塞电缆的需要，减少了无线电频率干扰，避免了烧毁、磨损或松动的电缆引起的潜在失火问题，减小了点火线圈和火花塞之间线路的电阻。因此，每个点火线圈都可以更小、更轻，使用更少的点火能量来点燃火花塞。

（6）冷却系统。

由于燃油与空气在气缸内燃烧，因此发动机部件温度升高，直接影响发动机性能和发动机零部件的使用寿命。冷却系统可以使发动机保持在有效的温度下运转，无论在何种驾驶条件下，该系统都旨在防止发动机过热和过冷。冷却系统包括散热器、风扇、水泵、水套、节温器等。它包括大循环和小循环，节温器是控制大循环和小循环路径的阀门，它决定了冷却液流向散热器的流量。

起动发动机后，发动机应尽快加热到有效的工作温度，并保持该温度而不过热。因此节温器保持关闭，冷却液在发动机缸体和气缸盖内进行小循环。当温度达到预定温度时，节温器打开，冷却液流过散热器，进行大循环。在很多维修案例中，当节温器出现故障而保持关闭时，冷却液因不能流向散热器而造成发动机过热；相反，如果节温器一直保持开启，发动机就会过冷。

（7）润滑系统。

发动机有许多相互接触的运动部件，当它们相互运动时会产生磨损。发动机润滑油在这些运动部件之间循环，可以防止金属间的接触面磨损。由于润滑后的零部件因摩擦力减小而易运动，因此摩擦造成的能量损失可以降到最低。润滑油可作为密封介质以防止泄漏，如气缸壁上的润滑油膜有助于活塞环的密封，从而提高了发动机的压缩压力。润滑油还可作为冷却剂。

第 29 课　汽车底盘介绍

汽车底盘由传动系统、转向系统、行驶系统和制动系统四部分组成，其作用是支承车身，接收发动机的动力，确保汽车正常行驶。

1. 传动系统

传动系统的作用是将发动机输出的动力传递给驱动轮，其主要包括离合器（只配备于手动变速器，用于暂时切断发动机与驱动轮的动力传递）、变速器、主减速器、差速器、驱动轮等。发动机和驱动轮的位置决定了该汽车是属于发动机前置前轮驱动、发动机前置后轮驱动、发动机中置后轮驱动、发动机后置后轮驱动还是发动机四轮驱动。

（1）离合器。

踩下离合器踏板时，离合器中断发动机与传动系统的动力传递。抬起离合器踏板时，从动盘与飞轮接合，开始动力传递。

（2）变速器。

变速器是传动系统的关键部件，可以较好地利用发动机输出的动力和转矩，从而挂高速挡实现高速、低转矩或者挂低速挡实现低速、高转矩来适应不同的驾驶条件。变速器主要有手动变速器和机械式自动变速器等。

① 手动变速器（MT）。MT可以通过手动改变换挡杆位置来改变变速器中的齿轮啮合位置。MT通常有五个前进挡，现在有些车辆已经有六个或七个前进挡。前进挡挡位数越多，换挡越顺畅，汽车的燃油经济性就越高。

② 自动变速器（AT）。AT是汽车上较复杂的机械部件，主要包括机械系统、液压系统和电控系统，它们协调工作。AT使用行星齿轮机构实现换挡。它根据加速踏板踩下去的程度和车速实现自动换挡。驾驶人只需操作加速踏板就可以控制车速。

2. 转向系统

转向系统的作用是在汽车行驶时为驾驶人提供方向控制。作为转向系统的核心，转向器类型较多，其中齿轮齿条式转向器和循环球式转向器应用较广泛。齿轮齿条式转向器主要应用于乘用车上；而循环球式转向器主要应用于商用车上，尤其是重型商用车。齿轮齿条式转向器体积小、质量轻、维修方便，可以给驾驶人提供更多反馈和更好的路感。循环球式转向器耐用，转向响应好，驾驶人的路感好。

动力转向系统是在机械转向系统的基础上增加一套助力装置来减轻驾驶人转动转向盘的力。依据所用能源的不同，动力转向系统主要分为液压助力转向系统和电控助力转向系统。

（1）液压助力转向系统。

液压助力转向系统主要由转向油罐、转向油泵、转向控制阀、转向液压缸和相应管路组成。转向控制阀为转向液压缸提供了与转向盘旋转运动相应的液压。转向液压缸将施加的液压转换为助力作用在转向器上，增大了驾驶人施加的转向力。当转向盘居中时，转向控制阀阀体不动，转向油泵输送的油液全部流回转向油罐。当转向盘转动时，转向控制阀相应地打开或关闭。一侧油液将直接流回转向油罐，另一侧油液将继续流入转向液压缸，在活塞两侧产生油液压力差，从而产生辅助力，使转向更容易。

（2）电控助力转向系统。

电控助力转向系统包括转矩传感器、速度传感器、ECU 和电动机，利用电动机为汽车提供方向控制。电控助力转向系统可以根据车速改变助力，使得转向盘在低速时更轻便，在高速时更稳定。电控助力转向系统正逐渐取代液压助力转向系统，成为汽车制造商设计的主流。由于装有电控助力转向系统的汽车不需要发动机提供动力，因此可以减少燃油消耗，从而提高汽车的动力性能。

3．行驶系统

行驶系统包括悬架、车架、车桥、车轮四部分。行驶系统的作用是连接车轮和车身，通过传动系统接收发动机的动力，迅速消除弹性元件的振动。

（1）悬架。

悬架是车架（或车身）与车桥（或车轮）弹性连接装置的总称，包括弹性元件、减振器和导向装置。悬架可分为独立悬架和非独立悬架两种，在现代汽车上广泛应用独立悬架。采用独立悬架的汽车，每个车轮都单独安装在车架上，有单独的弹性元件和减振器，提高了乘坐舒适性。

汽车悬架上的弹性元件主要有三种，分别是螺旋弹簧、钢板弹簧和扭杆弹簧。汽车上广泛应用螺旋弹簧，钢板弹簧主要用在大多数卡车和重型汽车上。当弹性元件被压缩后释放时，它会振动一段时间，该振动作用在汽车上将降低乘坐舒适性。因此，在汽车上配置减振器来吸收储存在弹性元件中的能量并缩短汽车振动的时间。在汽车悬架系统中广泛采用液力减振器。

（2）四轮定位。

四轮定位对于当今汽车设计来说是必不可少的。适当的车轮定位不仅对汽车的行驶安全控制起着至关重要的作用，而且对汽车的乘坐舒适性控制起着举足轻重的作用。四轮定位仪是一种用来检测车轮定位参数，并将其与汽车制造商提供的说明书里的标准值进行比较的精确测量装置，同时为使用者提供指令来调整相应的定位参数，以获得最佳转向性能并减少轮胎的磨损。

4．制动系统

制动系统是汽车上很重要的一个关系到安全的系统。它的功能是使车辆在尽可能短的距离内停下，这是通过将车辆的动能转化为散发到大气中的热能实现的。一般来说，每辆汽车都有两套独立的、不同功能的制动系统，一套是行车制动系统，一般用脚操作；另一套是驻车制动系统，一般用手操作。

随着汽车电子技术的不断进步，制动系统已经从过去的机械式逐渐转变成现在的电子式。电子制动系统为汽车提供了更高的安全性，同时通过确保驾驶人辅助功能（包括起步时自动释放制动和斜坡起步时的坡道驻车功能）提高了操作的便利性。

行车制动系统由供能装置、控制装置、传动装置和制动器四部分组成。其中，制动器是抑制汽车运动的装置，是汽车的决定性部件。它吸收运动部件的能量，并借助摩擦力使汽车减速。每个车轮上均装有制动器总成，其型式或为鼓式或为盘式。当驾驶人踩下制动踏板时，制动器在液压或气压的作用下执行制动操作。

(1) 鼓式制动器。

鼓式制动器有两个制动蹄，制动蹄固定在制动底板上，在液压缸的作用下内部扩张与制动鼓接触。当驾驶人踩下制动踏板时，制动主缸的两个活塞分别移动，迫使制动液经过制动管路流入各制动轮缸，使得鼓式制动器的制动蹄与制动鼓接触，产生摩擦力，迫使车轮速度降低或者停止运转。当驾驶人抬起制动踏板时，回位弹簧收缩，使制动蹄远离制动鼓，轮缸活塞移动使制动液流回制动主缸以重新补给制动储液罐。

(2) 盘式制动器。

液压驱动的盘式制动器广泛应用于乘用车制动系统中。目前汽车上使用的盘式制动器有两种：一种是固定钳盘式制动器，另一种是浮动钳盘式制动器。

鼓式制动器的主要缺点是摩擦区域几乎完全被衬片覆盖，因此大部分热量必须通过制动鼓传导到外部空气中冷却。由于盘式制动器暴露在空气中，因此其更容易将热量散发到空气中。与鼓式制动器相比，盘式制动器具有更好的抵制衰退的性能（热导致制动效率下降），这意味着盘式制动器可以长时间连续工作。盘式制动器具有更高的渐进制动效率。盘式制动器的内侧和外侧制动衬块的磨损相等，具有相对恒定的制动系数性能和较低的褪色敏感性。但是，盘式制动器也有一些缺点，其在重型商用车上使用时刹车片的使用寿命短，需要较高的购置成本和运行成本，容易产生制动噪声。

此外，鼓式制动器的零件比盘式制动器多，更难维修；但它的制造成本更低，而且易与紧急制动机构结合。由于鼓式制动器的有效制动摩擦面积大于盘式制动器，因此鼓式制动器具有更高的制动效率，这是重型商用车所必需的。大多数客车都使用盘式制动器，现在有些商用车也采用盘式制动器。

第 30 课　新能源汽车

新能源汽车发展提速是全球迈向碳中和方案的重要途径之一。新能源汽车相关技术在不断完善的同时，为这个行业带来了很多发展机遇。在我国，《新能源汽车产业发展规划（2021—2035 年）》旨在促进和激励我国新能源汽车产业高质量、可持续发展。该计划由国务院发布，列出了五项战略任务：提高技术创新能力，建设新型产业生态系统，推进产业整合，完善基础设施体系，深化开放合作。根据该计划，到 2035 年，纯电动汽车有望成为新车销售的主流，并且公共交通将全部电气化。

新能源汽车减少了对石油的依赖，利用相对便宜的电力资源。推动电动汽车市场发展的主要因素包括环境压力、能源安全、技术进步和政府政策的支持。在消费领域，优惠政策鼓励人们选择新能源汽车而非燃油汽车，这促进了新能源汽车市场的扩张，并促使产业链上的企业加大创新力度。

按照 2016 年 8 月 12 日工业和信息化部发布的《新能源汽车生产企业及产品准入管理规定（修订征求意见稿）》对新能源汽车的定义及技术指标有了更新的定义和更严苛的标准，新能源汽车是指采用新型动力系统，完全或主要依靠新型能源驱动的汽车，包括插电式混合动力（含增程式）汽车、纯电动汽车和燃料电池电动汽车等。相比于传统燃油汽车，新能源汽车采用非常规的车用燃料作为动力来源，或者使用常规的车用燃料，但采用新型车载动力装置。新能源汽车综合了车辆的动力控制和驱动方面的先进技术，是具有新技术和新结构的汽车。

1. 插电式混合动力（含增程式）汽车

插电式混合动力（含增程式）汽车是具有高容量电池的混合动力汽车，可以通过将电池插入电源插座或充电站进行充电。在典型的驾驶条件下，它们可以储存足够的电力来显著减少燃油的使用。

有两种基本的插电式混合动力配置。串联插电式混合动力汽车，也称增程式电动汽车，只有电动机驱动汽车行驶，汽油发动机只发电。直到蓄电池电量耗尽之前，它只能靠电运行。之后，汽油发动机发电，为电动机提供动力。对于短途旅行，这些车辆可能根本不用燃油。对于并联插电式混合动力汽车或混合插电式混合动力汽车，发动机和电动机都连接在车轮上，在大多数驾驶条件下驱动车辆。插电式混合动力（含增程式）汽车的动力驱动系统如图30.1所示。纯电动操作模式通常只在低速时发生。

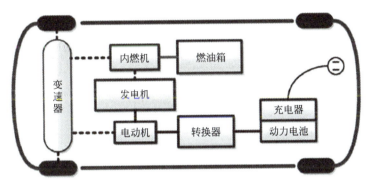

图 30.1 插电式混合动力（含增程式）汽车的动力驱动系统

一些插电式混合动力汽车拥有更高容量的蓄电池，续驶里程可以比其他汽车长。插电式混合动力汽车的燃油经济性对驾驶风格、驾驶条件和附件使用非常敏感。

2. 纯电动汽车

纯电动汽车有可充电蓄电池，没有汽油发动机。所有驱动车辆的能量都来自蓄电池组，蓄电池组由电网充电。纯电动汽车是零排放汽车，因为它们不会排放任何有害的尾气，而传统燃油汽车会造成空气污染。纯电动汽车的蓄电池组必须定期充电。最常见的充电方式是通过家用充电插头、公共充电站或定制的商业插座从国家电网充电。尽管充电时间受电网连接容量的限制，但通常被设计为一夜就能充满电。纯电动汽车所需的充电基础设施本身存在许多技术问题，需要开发低成本、可靠的移动解决方案。

3. 燃料电池电动汽车

燃料电池电动汽车可大幅度减少对燃油的依赖，并可降低导致气候变化的有害气体排放量。燃料电池电动汽车使用氢气而不是汽油运行，没有有害的尾气排放。要想与传统燃油汽车竞争，必须克服一些挑战，但它们的潜在好处是巨大的。燃料电池电动汽车看起来像传统燃油汽车，但使用了尖端技术。燃料电池电动汽车的核心是燃料电池堆。燃料电池堆将储存在汽车上的氢气和空气中的氧气转化为电能，为汽车的电动机提供动力。燃料电池电动汽车的主要组成如图30.2所示。

燃料电池是不用通过燃烧就能将化学能转化为电能的电化学装置。与蓄电池不同，只

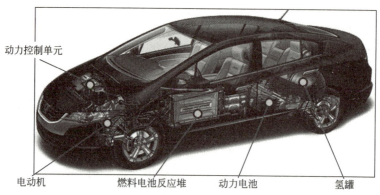

图 30.2　燃料电池电动汽车的主要组成

要有燃料供应，燃料电池就能持续发电。燃料的储存，特别是氢气的储存是有安全问题的，因为氢气易燃，并且很容易从压力容器中溢出。制造压力容器的材料也可能会出现耐久性问题。对于燃料电池驱动的车辆来说，氢气是一种有吸引力的燃料，但其运输和储存带来了挑战。使用氢气作为燃料的技术问题是，虽然氢气可以通过管道运输，但氢气容易泄漏，并会使一些用于管道、阀门等的金属脆化。

与纯电动汽车不同的是，燃料电池电动汽车可以像传统燃油汽车一样快速补充燃料，行驶时可以比其他类型的汽车安静、平稳、高效。燃料电池电动汽车被认为是未来电动汽车的发展方向。

近年来，中国新能源汽车行业的发展势头强劲，在全球汽车电动化中抢占了先机，并促进了国内汽车品牌的竞争优势。在政策支持和市场创新驱动的推动下，中国已经建立了一个相对完整的电动汽车产业链，涵盖了动力电池、电动机、电子控制以及整车制造和销售。在全球电气化和智能化浪潮中，中国汽车制造商必须加快技术创新，使汽车制造更加数字化和智能化。中国敞开怀抱欢迎外国新能源汽车制造商，他们发现有必要与中国同行合作或参与中国的新能源汽车供应链，以提高在全球市场的竞争力。

第 31 课　自动驾驶汽车

自动驾驶汽车是一种能够感知环境并在无人参与的情况下运行的车辆，也称自动驾驶车辆、无人驾驶车辆或机器人车辆。自动驾驶汽车不要求驾驶人在任何时候都控制车辆，甚至不要求驾驶人出现在车辆中。自动驾驶汽车可以去传统燃油汽车去的任何地方，做有经验的驾驶人做的所有事情。一辆完全自动驾驶的汽车有自我意识，能够自己作出决策。随着技术的不断成熟，自动驾驶汽车正在成为一种标准。

基于驾驶自动化系统能够执行动态驾驶任务的程度，根据在执行动态驾驶任务中的角色分配及有无设计运行范围，GB/T 40429—2021《汽车驾驶自动化分级》将驾驶自动化分为 0～5 级。

0 级驾驶自动化（应急辅助）不是无驾驶自动化，0 级驾驶自动化系统可感知环境，并提供信息或短暂介入车辆控制以辅助驾驶人避险。1 级驾驶自动化（部分驾驶）系统和 2 级驾驶自动化（组合驾驶辅助）系统和驾驶人共同执行全部动态驾驶任务，驾驶人监督系统的行为和执行适当的响应或操作。

对于3级驾驶自动化（有条件自动驾驶），动态驾驶任务后援用户以适当的方式执行接管。4级驾驶自动化（高度自动驾驶）系统发出介入请求时，用户可不作响应，系统具备自动达到最小风险状态的能力。

5级驾驶自动化（完全自动驾驶）系统发出介入请求时，用户可不作响应，系统具备自动达到最小风险状态的能力；5级驾驶自动化在车辆可行驶环境下没有设计运行范围的限制（商业和法规因素等限制除外）。5G的应用将实现5级驾驶自动化，使车辆不仅能够相互通信，还能与交通信号灯、交通标志甚至道路本身通信。

1. 自动驾驶汽车的工作原理

自动驾驶汽车依靠传感器、执行器、复杂的算法、机器学习系统和强大的处理器执行软件。自动驾驶汽车依据汽车不同部位的各种传感器创建和维护其周围环境地图。雷达传感器监测附近车辆的位置。摄像机探测交通信号灯，读取交通标志，跟踪其他车辆，寻找行人。激光雷达（激光探测及测距）从车辆周围反射光脉冲，以测量距离、检测道路边缘和识别车道标志。停车时，车轮上的超声波传感器检测路缘和其他车辆信息。复杂的算法处理输入信号并绘制路径，向执行器发送指令，执行器控制加速、制动和转向等。硬编码规则、避障算法、预测建模、对象识别帮助系统软件遵循交通规则和躲避障碍。

（1）自适应巡航控制系统。

自动驾驶汽车使用的一项车辆技术是自适应巡航控制系统。该系统能够自动调节车速，以确保与前方车辆保持安全距离；依赖汽车上使用的传感器获得的信息，允许汽车在感知到正在接近前方任何车辆时执行相关任务，如执行制动。处理相关任务信息后，适当的指令被发送到车辆执行器，执行器控制汽车的反应动作，如控制汽车转向、加速和制动。具有4级驾驶自动化的车辆能够对交通信号灯的信号和非车辆活动的信号作出响应。

（2）先进驾驶辅助系统。

先进驾驶辅助系统（ADAS）是在驾驶过程中帮助驾驶人的系统。首先，ADAS应增强汽车安全性，即增强道路安全性，ADAS的首要目标是在安全方面为驾驶人提供帮助，对驾驶人自己、其他车辆和行人交通参与者来说也是如此。其次，ADAS应提高驾驶舒适性。最后，也是最重要的，ADAS应改善经济和环境平衡。大多数交通事故都是人为失误造成的。ADAS是用以实现自动化、适应和增强车辆安全性和驾驶便利性的系统。ADAS被证明可以通过最大限度地减少人为错误来减少道路伤亡人数。ADAS的安全功能设计是通过提供提醒驾驶人潜在问题的技术或者通过实施安全措施和接管车辆的控制权来避免碰撞的。ADAS的自适应功能可以自动照明，提供自适应巡航控制，避免行人碰撞，提醒驾驶人注意其他汽车或危险，具有车道偏离警告功能，能够自动保持在车道中间行驶。

许多ADAS正处于新兴汽车技术的前沿。事实上，其中一些还没有定论。部分ADAS具有持久力，可以期待在未来汽车中看到。其他ADAS可能会失败并消失，或者被基本理念相同而更好实施的系统取代。

随着语音识别技术的深入应用，汽车工程师们也在寻求方法将其更好地整合到汽车系统中。当汽车研发者将语音识别视为复杂的人机界面的核心时，控制技术领域将发生颠覆性的改变。

2. 自动驾驶汽车的优势

与人类驾驶的汽车相比，自动驾驶汽车的一个潜在优势是可以提高道路安全性。车辆碰撞每年都会导致许多人伤亡，自动驾驶汽车可能会减少道路伤亡人数，因为其使用的软件系统与人类相比可能会更少犯错误。事故减少也可以减少交通拥堵，这是自动驾驶汽车具有的另一个潜在优势。自动驾驶汽车可以通过消除导致道路堵塞的人类行为来实现这一点。自动驾驶还有一个可能的优势，受年龄和残疾等因素的影响，不会开车的人可以使用自动驾驶汽车作为更方便的交通工具。此外，自动驾驶汽车可以消除驾驶疲劳，驾驶人能够在夜间行驶时睡觉。

3. 采用自动驾驶汽车技术的障碍

虽然 5 级驾驶自动化技术正在测试，但是离大众可以使用完全自动驾驶汽车还有很长的路要走。尽管自动驾驶汽车已经走过漫长的道路，但它还不是一种常见的交通工具，存在各种应用障碍。特别是制造一辆功能性自动驾驶汽车所需的技术成本非常高，导致汽车的最终成本也非常高。如果实现了雷达和激光雷达的大规模生产，那么它们的信号和频率可能会相互干扰。如遇路面上有雪、垃圾或油，无论何时覆盖车道标志和分隔带，对自动驾驶汽车来说都可能是一个重大挑战。此外，现在相信人工智能足够聪明，能够在生死攸关的情况（如当行人在急转弯后突然横穿马路）下作出与人类驾驶人一致的决定还为时过早。

第 32 课　如何撰写科技论文

科技论文是为了描述最新研究成果而写作并出版的论文报告。但这一简短的定义必须有一个前提条件限制，即科技论文必须以一定的形式写作并以一定的方式出版。

1. 题目

拟订科技论文题目时，作者应牢记：科技论文的题目将被成千上万的读者阅读，但会阅读整篇论文的只有少数几个读者，多数读者会通过原始期刊或二次文献（索引或摘要）阅读其题目。因此，题目中的每个词都应经过仔细推敲，词与词之间的逻辑关系也应认真处理。

好的题目用精炼的文字确切表达文章的内容。

科技论文的题目像是一个"标签"，而不是一个句子，所以它不必像句子一样具有主语、谓语。它确实比句子简单（或者至少通常比较短），但是也正因如此，词的排列顺序显得更为重要。

题目中每个词的含义和词序对阅读期刊目录中的标题来说都是非常重要的。而这一点对所有可能使用文献的人，包括通过二次文献查找科技论文的人（可能大多数读者如此）同样重要。因此，不仅题目本身应与科技论文相符，它的形式还应适合工程索引（EI）、科学引文索引（SCI）等科技文献检索系统。由于大部分索引和摘要服务系统都采用关键词分类法，因此作者确定科技论文题目时，最重要的是提供确切表达科技论文内容的关键词，也就是说，科技论文的题目用词应易理解、便于检索、突出重要内容。

2. 摘要

摘要是一篇科技论文（学位论文）简明、精确的概要。摘要的作用不是评价或解释科技论文，而是描述科技论文。摘要包括科技论文中问题或论点的简洁且精确的陈述、研究方法及设计思路的描述、主要发现及其意义、结论。摘要应包含能表达科技论文方法和内容的重要词汇（关键词），以方便检索，并使读者迅速、准确地了解科技论文的基本内容，据此确定是否与所需论文相关，进而决定是否要阅读全文。

摘要应写成完整的句子形式，而不能写成短语或电报式文字。一般来说，一篇科技论文的摘要应限制在200～250字，清楚地反映科技论文的内容。许多人只阅读原始期刊、EI、SCI或者其他二次出版物上的摘要。

摘要切忌提及科技论文中没有涉及的内容或结论。在摘要中不要引用与该论文有关的参考文献（极少数情况除外，如对以前发表方法的改进）。摘要不作为科技论文主体的一页与其一起计数和统计。

3. 引言

虽然题目和摘要都在科技论文的前面，但是一些有经验的作者往往在写好科技论文后写题目和摘要。写科技论文时，心里必须有（如未写在纸上）一个暂定的科技论文题目和提纲，还应考虑读者的水平，并据此确定需要具体定义或描述的术语及方法。

当然，正文的第一部分应是引言。引言的目的是向读者提供足够的背景知识及设计思路，使读者无须查阅与此研究课题有关的出版物就能够正确了解和评价科技论文中的研究结果。还应在引言中提出该项研究的理论基础。最重要的是，应简要、清楚地说明写该科技论文的目的，并审慎选定参考文献以获得重要的背景资料。

要想写出一篇好的引言，建议遵守下列规则。

（1）应尽可能清楚地提出研究问（课）题的性质和范围。

（2）为了适应读者的需要，应对有关文献进行评述。

（3）应介绍采用的研究方法和手段，如有必要还应说明理由。

（4）应介绍主要研究成果。

（5）应介绍由结果得出的结论。

4. 材料与方法

在"材料与方法"部分应提供详细的实验细节，主要目的是详细介绍实验方案，以使有能力的研究人员重复这个实验。许多（可能是大部分）读者可能会略过这部分，因为他们在引言中已经知道使用的一般方法或者对实验细节不感兴趣。但是，认真撰写这部分非常重要。因为科学方法的核心就是要求研究成果不仅要有科学价值，而且必须是能够实践的。为了判断研究成果能否再次实验成功，必须为其他人提供实验依据。不能重复的实验是没有意义的，能重复的实验必须能产生相同或相似的结果，否则科技论文的科学价值不大。

当科技论文被同行审阅时，审稿人会认真阅读这部分。如果怀疑实验能否再现，那么无论研究成果多么令人信服，这个审稿人都会建议退回稿件。

材料应包括准确的技术规格、质量、来源及制备方法等。通常还必须列出实验试样或试剂的有关化学性能及物理性能。

通常按时间顺序介绍方法。但很明显，相关方法应放在一起介绍，而不要总是拘泥于顺序平铺直叙。如果方法是新的（未发表过的），就应提供所需全部实验细节。如果一个实验方法已在正规期刊上发表过，那么只需给出参考文献即可。

5. 结果

科技论文的核心部分是数据，这部分称为结果。

通常应在"结果"部分全面描述实验，给出一个大致轮廓，但不要重复已经在"材料与方法"部分提到的实验细节，还应提供数据。

当然，提供数据并不是一件简单的事。直接将实验记录本上的数据抄到科技论文上的做法是极少有的。完成科技论文时，要选择具有代表性的数据，而不是罗列出所有数据。

因为结果是作者提供给世界的新知识，所以"结果"部分应写得清楚、简练。科技论文的前几（"引言""材料与方法"）部分告诉读者作者为什么及如何得到这些结果，而科技论文的后面（"讨论"）部分告诉读者这些结果的意义。很明显，因为整篇科技论文都是以"结果"为基础的，所以"结果"部分必须非常清楚地表述。

6. 讨论

与其他部分相比，"讨论"部分的内容更难以确定，因此它是最难写的部分。虽然许多科技论文中的数据正确且有根据，能够引起读者的兴趣，但"讨论"部分写得不好会使其被期刊编辑拒绝，甚至可能由"讨论"部分的解释说明使数据的真正含义变得模糊不清导致退稿。

"讨论"部分的基本特征如下。

（1）尽量叙述清楚"结果"部分显示的原理、相互关系和归纳性解释。谨记，好的"讨论"应该是对"结果"进行讨论，而不是扼要重述。

（2）要指出例外或有关不足之处，并应明确提出尚未解决的问题。千万不要冒风险去试图掩盖或捏造数据。

（3）应说明和解释结果与以前发表过的研究结果相符（或有差别）之处。

（4）大胆论述研究工作的理论意义及任何可能的实际应用。

（5）尽可能清晰地叙述结论。

（6）应简要叙述每个结论论据。

描述观察的事物之间的相互关系时，不需要得出一个广泛适用的结论。因为很难阐明全部真理，所以尽最大努力所能做到的就是像聚光灯一样照耀在真理的某部分，关于这部分的真理是由数据支持的。如果将数据外推到更大范围，那么会显得荒唐，此时甚至连数据支持的结论都可能受到怀疑。

当叙述真理的意义时，要尽可能简单。有时，最简单的语言可以表达最多的才智与学识；冗长累赘的语言或华丽的艺术辞藻只能表达肤浅的思想。

参 考 文 献

蔡安薇，崔永春，2009. 汽车专业英语［M］. 2版. 北京：北京理工大学出版社.
方梦之，毛忠明，2005. 英汉-汉英应用翻译教程［M］. 上海：上海外语教育出版社.
贺自强，1989. 机械工程专业英语［M］. 北京：北京理工大学出版社.
黄运尧，司徒忠，1997. 机械类专业英语阅读教程［M］. 北京：机械工业出版社.
蒋忠理，2018. 机电与数控专业英语［M］. 北京：机械工业出版社.
李洪涛，费维栋，2001. 材料科学与工程专业英语：修订版［M］. 哈尔滨：哈尔滨工业大学出版社.
李俊玲，罗永革，2001. 现代汽车专业英语［M］. 北京：北京理工大学出版社.
刘瑛，阎昱，2015. 材料成型及控制工程专业英语［M］. 北京：机械工业出版社.
刘振康，1988. 机械制造英语读本［M］. 北京：机械工业出版社.
马玉录，刘东学，2001. 机械设计制造及其自动化专业英语［M］. 北京：化学工业出版社.
施平，1996. 机电工程专业英语阅读：修订版［M］. 哈尔滨：哈尔滨工业大学出版社.
司徒忠，李璨，2001. 机械工程专业英语［M］. 武汉：武汉理工大学出版社.
粟利萍，2005. 汽车实用英语［M］. 北京：电子工业出版社.
汤彩萍，2005. 数控技术专业英语［M］. 北京：电子工业出版社.
唐国全，何小玲，2004. 科技英语论文报告写作［M］. 北京：北京航空航天大学出版社.
王斌，马进，2006. 中文版 AutoCAD 2006 实用教程［M］. 北京：清华大学出版社.
王锦俞，闵思鹏，2002. 图解英汉汽车技术词典［M］. 北京：机械工业出版社.
叶邦彦，陈统坚，2005. 机械工程英语［M］. 2版. 北京：机械工业出版社.
张芳，林良明，2001. 微机械的基本特征、关键技术及应用前景［J］. 传动技术，15（1）：25-33.
张崎，杨承先，2005. 现代机电专业英语［M］. 北京：清华大学出版社.
张晓黎，李海梅，2005. 塑料加工和模具专业英语［M］. 北京：化学工业出版社.
章跃，张国生，2002. 机械制造专业英语［M］. 北京：机械工业出版社.
钟似璇，2004. 英语科技论文写作与发表［M］. 天津：天津大学出版社.
周志雄，孙宗禹，2000. 机械设计制造及其自动化英语教程［M］. 长沙：湖南大学出版社.
ARKIN, 1998. Behaviour-based robotics［M］. Cambridge：MIT Press.
ARKIN, FUJITA, TAKAG, et al., 2003. An ethological and emotional basis for human-robot interaction［J］. Robotics and autonomous systems，42：191-201.
ASKLAND, PHULÉ, 2005. 材料科学与工程基础［M］. 北京：清华大学出版社.
DAY, 1998. How to Write & Publish a Scientific Paper［M］. 5th ed. Arizona：Oryx Press.
EPSTEIN, Millimeter-scale, MEMS gas turbine engines. Proceedings of ASME Turbo Expo 2003 Power for Land，Sea，and Air，June，16-19，2003［C］. atlanta，Georgia：ASME.
GRABOWSKI, 2007. The illustrated AutoCAD 2008 quick reference［M］. New York：Thomson Delmar Learning.
JEN, GUTIERREZ, EAPEN, et al., 2002. Investigation of heat pipe cooling in drilling application：part Ⅰ：preliminary numerical analysis and verification［J］. International journal of machine tools and manufacture：design，research and application，42（5）：643-652.
KALPAKJIAN, SCHMID, 2003. Manufacturing processes for engineering materials［M］. 4th ed. New Jersey：Prentice Hall，Pearson Education Inc.

KRUUSMAA,2003. Global navigation in dynamic environments using case-based reasoning [J]. Autonomous robots,14 (1):71-91.

LYSHEVSKI,2002. Mechatronic curriculum-retrospect and prospect [J]. Mechatronics,12 (2):195-205.

MILLER,2002. 3D production drafting and presentation with AutoCAD 2002 [M]. New Jersey:Pearson Education Inc.

TOMIZUKA,2002. Mechatronics:from the 20th to 21st century [J]. Control engineering practice,10 (8):877-886.

VASILIEV,1996. Heat pipe science and technology [J]. International journal of heat and mass transfer,39 (14):2977-2987.